ſ ıvı ..

⌐uɔıes, Florı.

lume 1 in Studies in the Natural Sciences
·ries from the Center for Theoretical Studies

ιportance of modern science and tech-
s often lost in a myriad of highly
ʑed articles and journals. In this
volume, seven distinguished scientists
ιeir own specialties in historical per-
·, describing the interrelation of sci-
research and technologies ranging
ιermonuclear energy to television,
eir impact on all levels of society.

ιaureate W. E. Lamb, Jr. describes the
.ion and state of the art of lasers and
ʝeen, also a Nobelist, outlines the de-
.ent of the science of superconduc-
V. K. Zworykin, inventor of many
·n components and of the electron
·e, examines the implications of
ι today's world.

IMPACT OF
BASIC RESEARCH
ON TECHNOLOGY

Studies in the Natural Sciences

A Series from the Center for Theoretical Studies
University of Miami, Coral Gables, Florida

Volume 1 — IMPACT OF BASIC RESEARCH ON TECHNOLOGY
Edited by Behram Kursunoglu and Arnold Perlmutter • 1973

Volume 2 — FUNDAMENTAL INTERACTIONS IN PHYSICS
Edited by Behram Kursunoglu, Arnold Perlmutter,
Steven M. Brown, Mou-Shan Chen, T. Patrick Coleman,
Werner Eissner, Joseph Hubbard, Chun-Chian Lu,
Stephan L. Mintz, and Mario Rasetti • 1973

Contributors to This Volume

John Bardeen
University of Illinois at Urbana-Champaign
Urbana, Illinois

Manfred A. Biondi
University of Pittsburgh
Pittsburgh, Pennsylvania

P. A. M. Dirac
Center for Theoretical Studies
University of Miami
Coral Gables, Florida

Willis E. Lamb
Yale University
New Haven, Connecticut

Edward G. Ramberg
RCA Research Laboratories
Princeton, New Jersey

Hermann Robl
U. S. Army Research Office
Durham, North Carolina

Arthur L. Schawlow
Stanford University
Stanford, California

Edward Teller
University of California
Berkeley, California

Vladimir K. Zworykin
RCA Research Laboratories
Princeton, New Jersey
and
Center for Theoretical Studies
University of Miami
Coral Gables, Florida

IMPACT OF
BASIC RESEARCH
ON TECHNOLOGY

Edited by

Behram Kursunoglu
and
Arnold Perlmutter

Center for Theoretical Studies
University of Miami
Coral Gables, Florida

PLENUM PRESS • NEW YORK-LONDON • 1973

Library of Congress Catalog Card Number 73-82141

ISBN 0-306-36901-X

© 1973 Plenum Press, New York
A Division of Plenum Publishing Corporation
227 West 17th Street, New York, N.Y. 10011

United Kingdom edition published by Plenum Press, London
A Division of Plenum Publishing Company, Ltd.
Davis House (4th Floor), 8 Scrubs Lane, Harlesden, London, NW10 6SE, England

PREFACE

The Center for Theoretical Studies has been fortunate from the start to be able to be host to some of the leading personalities of contemporary science. It thus seemed particularly appropriate to ask some of the Center members whose work has been crucial to the development of modern science and technology to trace the development of their own discoveries and to attempt to assess the ways in which their research has influenced other parts of society. The program was made possible by support of the Army Research Office - Durham.

The results of this project are partially represented by the seven essays which constitute this volume. While the work of each of the authors has had a substantial impact on science in some way, not all have resulted in applications which are of direct benefit to man and society. Such a statement is of course applicable to many areas of fundamental science, in particular some of the frontier areas of recent decades, such as astrophysics, elementary particles physics or molecular biology. Furthermore, in these areas we do not foresee tangible consequences in the near future. Nevertheless, the evolution of such frontier fields is reminiscent of the flowering of the arts and sciences during the Renaissance, whose subsequent spread throughout the world has enriched our lives immeasurably. Such accomplishments have not only helped to kindle our curiosity, but have brought a profound sense of satisfaction to man and to his self-fulfillment which are no less important than those branches of science which have already yielded practical fruits.

The essays in this book cover a wide range of subjects; what they have in common is the fundamental importance of the underlying science of each of them. It is interesting to note, in spite of the diversity of the topics, how many references have been made to each other and to several of the other classic discoveries of

v

modern science (all of this without any communication
among the authors). We believe that the essays give
considerable insight into the highly nonlinear process
which is attendant to all technological advances and in
particular, to how the latter depend so intricately on
the frontier areas of science (or as we are often fond
of saying, "useless science"). We hope that this volume
will throw some light on such questions and will provide
a modest demonstration, for both friend and foe of basic
research, that not only is science one of the fundamental bases for the survival of man, but that the
negative attitudes advanced by the opponents of science
are of only transitory influence and ought not to be
adopted by those of us who are seriously concerned with
the near and long-range future of mankind.

> Behram Kursunoglu
>
> Arnold Perlmutter
>
> The Editors

CONTENTS

INTRODUCTION

Hermann Robl

U. S. Army Research Office

Durham, North Carolina

New concepts of the physical world and the motivation of great scientists have long been fascinating subjects for educated laymen. Public interest in the management of research is a more recent phenomenon. As a result of the rapid increase in the level of funding during the last two decades, the benefit that our society derives from the support of research has been questioned, and government agencies have been forced to justify basic research programs in terms of technological applications.

In spite of many relevant publications there remain widespread misconceptions about the impact of basic research on modern technology. Such recurring misunderstandings stimulated the idea of asking a few eminent scientists to describe the historical development of their field of specialization and to show how advances in the understanding of fundamental phenomena have led to important applications. We thought that physics, and quantum mechanics in particular, would be ideal subjects for a demonstration of the relation between basic and applied research.

Due to various circumstances, the original plan had to be abandoned, but a grant by the U. S. Army Research Office in Durham, N. C. (ARO-D), to the Center for Theoretical Studies at the University of Miami resulted eventually in a series of essays which cover various topics from quantum mechanics to electron optics. While each one of the uniquely qualified authors has described a period in the evolution of a special field, their interest

in and concern with technological applications varies
considerably. For this reason, some additional remarks
about the relationship between basic research and technology, and especially about the support of basic research
by mission oriented agencies in the U. S. appear to be
appropriate. In particular, we will refer to the support
program of ARO-D which is funding basic research projects
at educational institutions and industrial research centers, and we wish to explain some objectives of basic
research in government laboratories, so called "in-house"
basic research. In line with the scope of this volume,
all specific examples are drawn from the field of physics.

Two diametrically opposed misconceptions about the
contribution of basic research to the solution of technological problems are quite common. One of them is based
on the contention that all physical phenomena which may
have an impact on technology are already well understood.
From this point of view future technological progress
does not depend on more fundamental investigations but
only on applied research and inventive skill. At the
other extreme, one can find the mistaken belief that,
with a high degree of probability, any scientific investigation will lead directly to some future applications.

Actually, technological innovations are usually not
based on a single discovery, but on the integration of a
large number of individual research efforts. It is not
difficult, however, to provide some examples of technological applications of basic research during the last
few years, and one can identify technological barriers
which retard further progress until major scientific advances can be achieved. Mission oriented agencies aware
of the existence of specific technological problems are
obviously in the best position to orient basic research
programs toward the solution of these problems through
selection of research proposals and by placing emphasis
on relevant research areas.

Individual basic research projects funded by mission
oriented agencies are often indistinguishable from those
supported by the National Science Foundation. One has
to compare the distribution of research projects in scientific areas to see the difference between the research
program of the National Science Foundation and those of
mission oriented agencies. The support of basic research
by mission oriented agencies satisfies the needs of both
the sponsors and educational institutions, even if their
objectives are different.

There are reasons to believe that the most favorable condition for simultaneous progress of science and technology is the coincidence in challenge of scientific problems, or the interest in unexplained phenomena, and the recognition of potential applications by those who have the resources for the support of relevant applied research and development programs. The rapid growth of solid state physics coupled with corresponding advances in solid state electronics, and the development of a great variety of practical devices, are among the most convincing examples in support of our assertion. Similar developments have occurred in recent years in the fields of optics, plasma physics, and in atomic and molecular physics.

A brief recollection of the spectacular renaissance of optics which began in 1960 may be of interest. When the writer attended a conference on coherence properties of electromagnetic radiation which was held at the University of Rochester on 27-29 June 1960, it was obvious that an experimental demonstration of light amplification by stimulated emission was imminent. It was easy to predict that the spectral and directional characteristics of an optical maser would be ideally suited for a variety of applications. In a report dated 11 July 1960, the writer pointed out that "besides optical communication over very large distances, there are at least two possibilities which are of interest. . .High resolution radar and illumination of small targets for guided missiles." Following a technical discussion of these applications, including numerical estimates, we recommended that a basic research program be initiated "pertaining to the generation, modulation, and detection of coherent infrared and visible radiation," and that "high priority be assigned to this program."

Sufficient funds were obtained for several research projects at universities and industrial laboratories which in the following years made significant contributions in the field of gas lasers, glass lasers, and organic dye lasers. Information on the energy levels of rare gases, and rare earth ions which had been generated at The Johns Hopkins University several years earlier under ARO-D support, immediately became very useful. While several experimental projects were in progress, it was recognized that a deeper understanding of coherence properties and fluctuations of electromagnetic fields was needed, and support of relevant theoretical investigations was initiated at the University of Rochester, and later at the University of Southern California.

At the same time, the exploration by several investigators of nonlinear optical phenomena, especially the generation of harmonics of a fundamental optical frequency, parametric amplification and frequency conversion, was encouraged and supported by the Joint Services Electronics Program. All this work might have been done eventually without the support of mission oriented agencies, but there is no question that these agencies have strongly accelerated the rate of progress.

Prior to the initial investment of funds in laser physics by the Department of Defense, a similar impetus was given to atomic and molecular physics, especially to the investigation of atomic and molecular collision processes. This happened in view of great concern about the lack of information on phenomena in the upper atmosphere subsequent to the explosion of a nuclear weapon. In order to predict the attenuation of radar signals for the tracking both of offensive and defensive missiles it was necessary to measure cross sections and reaction rates of a large number of processes which are responsible for the production and removal of free electrons, and to determine electron collision rates. In present computer programs for the calculation of electron densities subsequent to a nuclear burst in the upper atmosphere, several hundred important reactions are taken into account, but 15 years ago many of the relevant cross sections and reaction rates were either unknown or too uncertain for reliable predictions. After the nuclear test ban treaty, the demand for laboratory measurements and for advances in the theory of atomic and molecular collisions became even more urgent. Again, the rate of progress in these fields was strongly affected by the Department of Defense.

Research administrators in mission oriented agencies are continuously reviewing progress in their field of interest in order to assess the potential contribution of scientific developments to the solution of current problems or their relevance to predictable future requirements. The orientation of the basic research programs of the mission oriented agencies depends strongly on these judgments. One should not assume, however, that every basic research project that is being supported by a mission oriented agency will contribute directly and immediately to applications and the solution of technological problems. As we have pointed out above a particular technological advance is usually not based on a single discovery or scientific investigation, but rather on the collective contribution of a large number of individual research efforts. A basic research program

may appear inefficient if a comparison is made of the
number of research projects funded, and the final applications. On the other hand, the investment in basic research is usually very modest compared with the cost of
hardware. All these factors should be taken into account
when the cost-effectiveness of a basic research program
is being questioned.

A discussion of the controversial issue of the "value" of technological progress is outside the scope of
this introduction, but we would like to add a few remarks
about the strong correlation of military and civilian
applications of basic research. For this purpose, consider the contribution of some selected subfields of
"classical" and "modern" physics to major areas of technology, as shown in a table at the end of this section.
Note that some potential rather than actual contributions
have been included, e.g., plasma physics with respect to
energy production. Except for the last two columns which
are concerned exclusively with military applications,
every technological application has civilian and military
implications, including those in the field of medicine.
For example, surveillance encompasses radar for air-traf-
fic control, as well as night vision devices for military
operations. As a general rule, any basic research in
physics which has a potential for civilian applications
can also lead to military applications, and conversely,
any basic research supported by the Department of Defense
may be expected to contribute to the welfare of our society beyond security in the military sense.

Unfortunately, it is very difficult to measure the
impact of basic research on technology. There is lack of
information on the actual use of basic research results,
because design engineers and manufacturers are usually
not concerned with the origin of basic ideas and principles. Likewise, it is not easy to determine the effects
of basic research on lead-time in development, production,
and the balance of payments in foreign trade, or the risk
of a technological surprise due to the military classification of a scientific discovery in a foreign country.
Congressmen may not be impressed by a few selected case
studies, and arguments based on possible consequences
of the neglect of basic research in the past are usually
oversimplified since alternatives tend to be ignored.
Most distressing is the fact that final decisions about
major development projects are influenced more strongly
by political and economic pressures than by objective
scientific considerations. Furthermore, some problems
such as those found in ballistic missile defense are so

extremely complex that even experts may disagree about
the feasibility of a systems concept.

Another matter which has attracted public attention
in recent years is the organization of those government
agencies which are concerned with the support of basic
research. The question is whether a centralized or mul-
tiple-agency mode of operation is more efficient. In
general, research administrators in these agencies and
scientists who have been supported by them agree that our
multiple agency system offers significant advantages.
We know that referees frequently disagree about the merit of research proposals, especially if they are faced
with highly original but untested ideas, or if a research
problem is very difficult. The use of independent groups
of referees reduces the risk of serious mistakes, but if
a research proposal could be submitted only to a single
agency, a decision about the support of this project could
be final. The major drawback of the present system is
the workload imposed on independent groups of referees,
which must be regarded as a lesser evil. Furthermore,
the critical evaluation of research proposals and an intensive review of ongoing research projects, as well as
personal contacts with leading experts in various fields
of science, enable research administrators to assess the
impact of scientific progress on future technology in
their fields of responsibility. The effective exchange
of information between scientists who are concerned with
basic research and those who are engaged in applied research, between scientists in educational institutions
and in government laboratories, which depends on the in-
itiative of mission oriented agencies, could probably
not be effected by a single super-agency.

So far we have discussed problems which are typical
for the support of basic research by mission oriented
agencies at educational institutions and industrial research centers. One may ask, why should basic research
also be performed in government laboratories which have
primary responsibilities for applied research, testing,
evaluation, and procurement of material. In particular,
one may ask "What is the role of basic research in laboratories of the Department of Defense, is it designed
to complement other basic research programs, or is it essential for the primary functions of these laboratories?"
In the opinion of the writer, which does not necessarily
reflect official policies, "in-house" basic research offers several significant advantages. First, as experience has shown, the competence of basic research oriented

scientists in government laboratories provides a quick
reaction capability for the solution of problems which
arise in in-house applied research and development. Second, it should be recognized that basic research oriented
scientists in the government laboratories are used efficiently for the evaluation and interpretation of highly
specialized information in professional journals for non-
specialists and design engineers. At the same time, they
are able to assist in the evaluation of research proposals submitted by educational institutions and private
laboratories with regard to scientific merit and mission
relevance. Finally, these scientists are fully qualified
to explain material requirements and research priorities
to other scientists in the academic community. But in
order to attract competent scientists for these assign-
ments, to maintain their qualifications, and to help
establish their reputation in the scientific community,
management must provide an opportunity for them to engage
in basic scientific investigations.

Although some emphasis has been placed in this introduction on technological applications of basic research,
the writer is well aware of its cultural value. The intellectual satisfaction derived from an understanding of
the physical world can best be compared with the appreciation of art, and like art it is not or should not be
the privilege of a small group of professionals. The
arts, science, and philosophy have been driving forces
in the history of civilization, and any nation which
strives to claim greatness has an obligation to uphold
their values.

Selected Relevant Publications

The Administration of Government Supported Research at
Universities, Bureau of the Budget, U. S. Government
Printing Office, 1966

National Science Policies of the U.S.A.
Origins, Developments and Present Status
UNESCO, 1968

TRACES, Technology in Retrospect and Critical Events in
Science. A report prepared by the IIT Research Institute
for the National Science Foundation. 2 Volumes, 1968/69

Science and Technology: Tools for Progress
The Report of the President's Task Force on Science Policy
U. S. Government Printing Office, 1970

The Physical Sciences
Report of the National Science Board, Submitted to the
Congress, U. S. Government Printing Office, 1970

Alvin M. Weinberg
In Defense of Science
Science, Vol. 167, p. 141, 1970

Philip Handler
The Federal Government and the Scientific Community
Science, Vol. 171, p. 144, 1971

Rodney W. Nichols
Mission-Oriented R&D
Science, Vol. 172, p. 29, 1971

	Medicine	Energy	Communications	Computers	Materials	Surveillance	Aviation	Ballistic Missile Defense	Nuclear Weapons
Fluid Dynamics							X	X	X
Plasma Physics		X	X						X
Nuclear Physics		X							X
Atomic and Molecular Physics					X			X	
Solid State Physics			X	X	X	X			
Quantum Electronics			X			X			
Electron Optics	X		X		X	X			

P. A. M. DIRAC

SOME OF THE EARLY DEVELOPMENTS OF QUANTUM THEORY

P. A. M. Dirac

Center for Theoretical Studies

University of Miami, Coral Gables, Florida

I would like to discuss my scientific work in a general and nontechnical way. I will try to impart some idea of the feelings of a research worker when he is hot on the trail and has hopes of attaining some important result which will have a profound influence on the development of physics. One might think that a good research worker in this situation would review the situation quite calmly and unemotionally and with a completely logical mind, and proceed to develop whatever ideas he has in an entirely rational way. But this is far from being the case. The research worker is only human, and if he has great hopes he also has great fears. I don't suppose one can ever have great hopes without their being combined with great fears. And as a result his course of action is very much disturbed. He is not able to fix his attention on the correct logical line of development.

I shall be relating mainly my own experience in this connection, but from talks which I have had with other physicists, some of them very eminent, I feel that what I have to say holds pretty generally and one can accept it as a general rule applying to all research workers who are concerned with the foundations of physical theory. They are influenced by their fears to quite a dominating extent.

I expect it was similar fears which applied to other cases where we don't have any direct evidence of what happened. In this connection I would like to refer particularly to Lorentz. Any one of you who has studied

relativity must surely have wondered why it was that
Lorentz succeeded in getting correctly all the basic
equations needed to establish the relativity of space and
time, but just wasn't able to make the final step establishing relativity. He did all the hard work - all
the really necessary mathematics - but he wasn't able
to go beyond that and you will ask yourself, why?

I think he must have been held back by fears. Some
kind of inhibition. He was really afraid to venture into
entirely new ground, to question ideas which had been
accepted from time immemorial. He preferred to stay on
the solid ground of his mathematics. So long as he
stayed there his position was unassailable. If he had
gone further, he wouldn't have known what criticism he
might have run into. It was the desire to stay on
perfectly safe ground which I presume was dominating him.

It needed several years and the boldness of Einstein
to make the necessary step forward and say that time and
space are connected. What seems to us nowadays a very
small step forward was very difficult for the people of
those days.

This is all just conjecture, of course, but I feel
that it must correspond rather closely to the facts. I
don't see any other explanation of how one can get so
near to a great discovery and yet fail at the last and
rather small step.

Let us pass on to consider the development of quantum
mechanics. This started with a brilliant idea of Heisenberg. Heisenberg's idea was that one should try to
construct a theory in terms of quantities which are provided by experiment, rather than building it up, as people
had done previously, from an atomic model which involved
a lot of quantities which could not be observed. By this
brilliant idea Heisenberg really started a new philosophy,
a philosophy that physics - physical theory - should
keep close to the experimentally obtained data and should
not depart into the use of quantities which are only very
remotely connected with observation.

This was a wonderful idea of Heisenberg's, and in
putting together the various experimentally provided
data concerned with atomic spectra he was led to matrices,
and then was led to consider that matrices represent the
physical variables occurring in an atom - physical
variables like the positions and velocities of the

electrons. Now Heisenberg had not proceeded very far
with this idea before he noticed that it would lead to
his physical quantities not satisfying the commutative
law of multiplication. Two such quantities A and B
would usually be such that A multiplied by B is different
from B multiplied by A.

Now when Heisenberg noticed that he was really
scared. It was such a foreign idea. Physicists right
from the earliest times had always thought of the variables that they were using as quantities satisfying the
ordinary laws of algebra. It was quite inconceivable that
two physical things when multiplied in one order should
not give the same result as when multiplied in the other
order. It was thus most disturbing to Heisenberg. He
was afraid this was a fundamental blemish in his theory
and that probably the whole beautiful idea would have
to be given up.

I received an early copy of Heisenberg's first work
a little before publication and I studied it for a while
and within a week or two I saw that the noncommutation
was really the dominant characteristic of Heisenberg's
new theory. It was really more important than Heisenberg's idea of building up the theory in terms of
quantities closely connected with experimental results.
So I was led to concentrate on the idea of noncommutation
and to see how the ordinary dynamics which people had
been using up till then should be modified to bring it in.

At this stage, you see, I had an advantage over
Heisenberg, because I did not have his fears. I wasn't
afraid if Heisenberg's theory would collapse. It wouldn't
have affected me like it would have affected Heisenberg.
It wouldn't have meant to me that I would have had to
begin again right from the beginning.

I think it is a general rule that the originator
of a new idea is not the most suitable person to develop
it, because his fears of something going wrong are really
too strong and prevent his looking at the method from
a purely detached point of view in the way that he ought
to.

I had this advantage over Heisenberg. I also had
other great advantages. I was a research student at
that time with no other duties except research. I can
thank the fact that I was born at just the right age. A
few years older or younger and I would have missed that

opportunity. But everything seemed to be in my favor.

Also, with regard to the problem of modifying the
ordinary dynamics to bring in noncommutation, I had got
used to the earlier theory of Bohr and Sommerfeld - the
theory of atomic orbits - and this theory had been found
to be closely connected with a form of mechanics due to
Hamilton, which he had discovered nearly a hundred years
previously. It turned out that Hamilton's form of dynamics was just the form which was most suitable for
bringing in noncommutation, and it was not a very difficult problem to work out how these two ideas should
be fitted together.

I was working on this subject quite independently
from Heisenberg after getting his initial idea. Heisenberg continued to work on it. He was collaborating with
other people in Göttingen. There was his Professor,
Born, and another young research student, Jordan, in
particular. I expect they were a great help to him in
overcoming his fears. The result was that the Göttingen
School also made rapid progress in developing the basic
ideas of quantum mechanics. We published our work independently at about the same time. If you look up these
early papers you will see that there is quite a difference
in our styles, because in my work the noncommutation was
the dominant idea. With the Göttingen School, the dominant idea was the use of quantities closely connected
with experimental results and the noncommutation appeared
as secondary and derived. But still, with these different points of view, there wasn't any real discrepancy
and we both got the same essential results.

There was another form of quantum mechanics which
was discovered quite independently by Schrödinger. He
was following some different ideas and had his own difficulties. His ideas were based on a remarkable connection between waves and particles which had been discovered
a little earlier by de Broglie. This connection of
de Broglie's was very beautiful mathematically and was
in agreement with the theory of relativity. It was very
mysterious, but because of its mathematical beauty one
felt that there must be some deep connection between the
waves and the particles illustrated by this mathematics.

De Broglie's ideas applied only to free electrons
and Schrödinger was faced with the problem of modifying
de Broglie's equation to make it apply to an electron
moving in a field, in particular to make it apply to
electrons in atoms. After working on this for some time,

Schrödinger was able to grasp at an equation - a very
neat and beautiful equation, which seemed to be alright
from a general point of view.

Of course, it was necessary then to apply it, to
see if it would work in practice. He applied it to the
problem of the electron in the hydrogen atom and worked
out the spectrum of hydrogen. The result that he got
was not in agreement with experiment. That was most
disappointing to Schrödinger. It was an example of a
research worker who is hot on the trail and finding all
his worst fears realized. A theory which was so beautiful,
so promising, just didn't work out in practice.

What did Schrödinger do? He was most unhappy. He
abandoned the thing for some months, as he told me. And
then, afterwards, when he had recovered from his depression,
somewhat, he went back to this work again and noticed that
if he applied his ideas with less accuracy - not taking
into account effects due to the relativistic motion of
the electron - with this lesser accuracy, his theory
agreed with observation. He then published his work with
this diminished accuracy and was able to establish that
his theory was in agreement with observation.

He set up in that way an alternative form of quantum
mechanics. People who were working on the subject soon
found that it was basically equivalent to the form which
had been originated by Heisenberg. They were just two
aspects of the same theory, which is our present quantum
mechanics.

Schrödinger had really been too timid in giving up
his first relativistic equation which was not in agreement with the observations of the hydrogen spectrum.
This equation was rediscovered a little later by Klein
and Gordon and they published their work in spite of its
disagreeing with observation. The reason why Schrödinger's
original equation did not fit was because it did not take
into account the spin of the electron. The spin of the
electron was not established at that time. There were
some hints of spin provided by experiment, but they were
very vague. Probably Schrödinger himself did not know
about them.

Klein and Gordon published the relativistic equation,
which was really just the same as the equation which
Schrödinger had discovered previously. The only contribution of Klein and Gordon in this respect was that they

were sufficiently bold not to be perturbed by the lack
of agreement of the equation with observations. The
result is that the equation is now known as the Klein-
Gordon equation, in spite of having been discovered a
year or two earlier by Schrödinger. This equation is
of some value for describing particles without spin -
certain kinds of mesons - but does not apply at all to
the electron.

That is the way in which quantum mechanics was
started. We had a quite definite mathematical theory
at the beginning and people were led slowly to find the
proper interpretation for the equations. It had to be
a statistical interpretation. Many people worked on
this. The problems were not really so difficult when
once the starting point had been firmly established.

Serious difficulties did not arise until one ex-
amined the corrections needed in the theory by relativity.
As I noted, Shrödinger's equation was valid only to the
approximation in which one neglects these corrections.
If one tried to work with the relativistic equation that
bears the name of Klein and Gordon, one not only had
disagreement with observation, but one also had disagreement with the logical interpretation of the equation.
Applying the usual rules which had been generally established for quantum mechanics, it would seem that the
Klein-Gordon equation would lead one to negative
probabilities, which are of course quite absurd.

There was some modification needed in the Klein-
Gordon equation. I puzzled over it for a time and
eventually I was able to think of another equation which
would get over the logical difficulties of the negative
probabilities. I soon saw that this new equation gave
correctly the spin and the magnetic moment of the electron.
That was all very satisfactory.

The question then arose, would it satisfactorily
explain the spectrum of hydrogen? I worked that out,
taking account only the first order of accuracy for the
relativistic corrections, and I got agreement with
experiment to this first order. Well the natural thing
to do then would be to examine the higher orders and see
whether they would also give agreement with experiment.
But I didn't do it, simply because I was scared. I was
afraid that maybe they wouldn't come out right. Perhaps
the whole basis of the idea would have to be abandoned
if it should turn out that it wasn't right to the higher

orders and I just couldn't face that prospect. So I
hastily wrote up a paper giving the first order of approximation and showing that to that accuracy, at any
rate, we had agreement between the theory and observation.
In that way I was consolidating a limited amount of
success, that would be something which one could stand
on independently of what the future would hold. One
very much fears the need for some consolidated success
under circumstances like that, and I was in a great
hurry to get this first approximation published before
anything could happen which might just knock the whole
thing on the head.

It was left to Darwin to fill up the gap that I
left. Darwin, you see, was able to approach the subject without having my fears. He made the necessary
calculations to all orders of accuracy and found that
they did fit, and I was very relieved to hear about it.

If you look up my first work on that subject (I
don't know if anybody does that nowadays except the
historians of science), there is one thing which you can
hardly fail to notice. There is an expression which I
wrote down which is roughly like this (I am just putting
down the essential points):

$$\frac{w^2}{c^2} + p_1^2 + p_2^2 + p_3^2 \ .$$

Now when you look at that, if you know anything at all,
you will say that that is wrong. There should be minus
signs here (before the p's). So you will conclude that
there was a misprint in the paper. But it was a very
prominent misprint and you will perhaps wonder how I
could have been so careless to overlook it.

I was a careful proof-reader in those days.

Well the explanation of that is that it is not
really a misprint, but the appearance of that equation
is again the expression of a fear. This work was done
in the 1920's when the whole idea of relativity was still
quite young. It didn't come to make a splash in the
scientific world until after the end of the first world
war and then it did make a very big splash. Everyone was
talking about relativity, not only the scientists, but
the philosophers and the writers of columns in the newspapers. I don't think there has been any other occasion
in the history of science when an idea has so much caught
the public interest as relativity did in those early days,

starting from the relaxation which occurred with the
ending of a very serious war.

Now the basic idea of relativity was a symmetry
between space and time. But this symmetry is not quite
perfect symmetry. In order to make it perfect, one has
to change the signs in some of the equations. One can
bring about that necessary change in sign by introducing
the root of minus one ($\sqrt{-1}$) into certain physical quantities (wherever we have a four-vector we have to introduce
the root of minus 1 in certain coordinates). Referring
to quantities which have been doctored in this way, one
has complete symmetry between space and time. Now the
early workers in relativistic theory were very much impressed by the symmetry between space and time and just
wanted to cling to it - to hold onto it at all costs.
And so they frequently used this notation containing the
root of minus one, just to bring in the complete symmetry.
The result was expressions like the one above. This
notation was quite common (I see in my early notes that
I was using it all the time). It was so common that
people didn't bother to mention it - whenever they used
it in a paper they let it be understood. One could see
from the signs in the expression whether the root of
minus 1 should be inserted in the basic variables or
not, and there was no need to waste time explaining it.
So what looks like a misprint nowadays, when people
no longer feel the need to cling to the symmetry of space
and time, was not a mistake, but was a historical consequence of the way relativity developed.

Now how did the further development of quantum theory
go after this stage? We had a relativistic equation
which worked. It gave agreement with experiment to high
accuracy for the simple example of the hydrogen atom.
It wasn't long before a new difficulty appeared, namely,
working with this equation, one found that the electron
had states of negative energy. For a particle to be in
a state of negative energy, of course, is something
which appears quite impossible. From an experimental
point of view, it is certainly never observed. So it
would seem that one had gotten over one difficulty only
to plunge into another.

It frequently happens with the development of science
that when one gets over one difficulty, one is immedi-
atly faced with a newer difficulty, and you might at
first sight think that no real progress has been made.
But real progress is made, because the new difficulty

is more remote than the previous one. If one looks
into things more closely, one usually sees that the new
difficulty was really there all the time. It was just
hidden previously and swamped by a more crude difficulty,
and when the cruder difficulty is explained away, people
focus their attention on the new difficulty.

When this new difficulty of the negative energy
states appeared it was an example of a difficulty that
wasn't really new; it was there all the time. In any
relativistic theory this difficulty occurs - even in the
old classical theory of Lorentz. But it didn't matter
under those conditions, because an electron could then
never jump into one of the states of negative energy.
There was continuity, which prohibited such jumps. However, with the new Quantum Theory such jumps could occur
and the difficulty could not be ignored in the way in
which it had been previously.

Well, I found that it wasn't really very hard to
see a way out of this difficulty. The idea was suggested
by the chemical theory of valency, in which one is used
to the idea of electrons in an atom forming closed shells
which do not contribute at all to the valency. One gets
a contribution from an electron outside closed shells
and also a possible contribution coming from an incomplete
shell or a hole in a closed shell.

One could apply the same idea to the negative energy
states and assume that normally all the negative energy
states are filled up with electrons, just in the way in
which the closed shells in the chemical atom are filled
up. In that way an ordinary positive energy electron
would not be able to jump into a state of negative
energy. However, one would expect that under certain
conditions there might be a hole in the negative energy
states and one had to get an interpretation for these
holes.

One can see at once that such a hole will appear
as a particle. It will be a particle with a positive
charge and a positive mass. Now right from the start
when I had this idea, it seemed to me that there would
be symmetry between the holes and the electrons and
therefore the holes must have the same mass as the
electrons. How could one then interpret the holes?
They would be particles of positive charge. The only
particles of positive charge known at that time were
protons. For decades physicists had been building up
their theory of matter entirely in terms of electrons

and protons. They were quite satisfied to have just
these two basic particles. The electrons carry the negative charge, the protons carry the positive charge.
That was all that was needed. Rutherford had put forward some tentative ideas that there just might be a
third particle - a neutron. That was just a speculation
which people talked about occasionally, but nobody took
it really very seriously.

On this basis that the only particles in nature are
electrons and protons, it seemed to me that the holes
would have to be the protons. And that was a great worry,
because the protons have a very different mass from the
electrons. They are very much heavier. How could one
explain this difference in mass?

I searched about for some time for some cause that
would explain it and hoped that maybe the Coulomb
force between the electrons might lead to some relationship between all the electrons in the negative energy
states which would lead to a difference in mass, though
I just couldn't see how it could come about. But still
I thought there might be something in the basic idea and
so I published it as a theory of electrons and protons,
and left it quite unexplained how the protons could
have such a different mass from the electrons.

This idea was seized upon by Herman Weyl. He said
boldly that the holes had to have the same mass as the
electrons. Now Weyl was a mathematician. He wasn't
a physicist at all. He was just concerned with the
mathematical consequences of an idea, working out what
can be deduced from the various symmetries. And this
mathematical approach led directly to the conclusion that
the holes would have to have the same mass as the electrons. Weyl just published a blunt statement that the
holes would have to have the same mass as the electrons
and did not make any comments on the physical implications of this assertion. Perhaps he didn't really
care what the physical implications were. He was just
concerned with getting the mathematics consistent.

At this stage in the course of development of the
theory, Oppenheimer made a contribution. Oppenheimer
accepted Weyl's conclusion that the holes had to have
the same mass as the electrons and faced the physical
reality that the holes were not observed in practice.
Oppenheimer just said that there was some reason, which
we don't understand, why the holes are never observed.
He agreed that the holes couldn't have anything to do

with protons, so there had to be some mysterious reason
why they did not occur in nature.

Well, Oppenheimer was really very close to the
mark with this hypothesis. The reason why the holes
were not observed was simply that the experimental
people had not looked for them in the right place, or
if they had looked, they hadn't recognized what they
saw.

I can remember in these early days, even somewhat
before this theory of electrons and protons, when talking with people who were working in the Cavendish and
were observing tracks of particles in a magnetic field,
they said that they sometimes observed an electron
going into the source. They treated these occurrences
as coincidences. Nobody thought it worthwhile to look
more closely into them. The idea of there being a new
particle coming out from the source, instead of an
ordinary electron going into the source, was completely
foreign to the accepted mode of thought of those days.
I don't think anybody had the remotest suspicion of such
a possibility. They had the evidence before their eyes
for these new particles with positive charge and the
same mass as the electrons, but they were just unable to
appreciate what they saw.

It needed some years of development by the experi-
menters before the fact of the existence of positrons
was established. Blackett was really the first to obtain
hard evidence for the existence of a positron but he
was afraid to publish it. He wanted confirmation. He
was really over-cautious. It was left to Anderson to
first publish the evidence for the existence of a positron and to scoop the credit for the discovery of the
positron.

When one thinks back to these days, one finds that
it is really remarkable how unwilling people were to
postulate a new particle. This applies both to theoretical and experimental workers. It seems that they
would look for any explanation rather than postulate a
new particle. It needed the most obvious and unassailable evidence to be presented before them before they
were reluctantly forced to postulate a new particle.
The climate has completely changed since those early
days. New particles are now being postulated and proposed continually, in large numbers. There are a
hundred or more in current use now-a-days. People are

only too keen to publish evidence for a new particle,
whether this evidence comes from experiment or from some
ill-established theoretical idea.

It was a very difficult first step to accept the
positron. That was followed closely by the discovery of
the neutron, confirming Rutherford's hypothesis of several years earlier, and then the neutrino and various
mesons were afterwards discovered.

Those were the early days in which the basis for
quantum mechanics was established. The foundations
were laid for a theory, which has been found to be a
good theory for explaining all atomic events when one
doesn't inquire into phenomena involving distances
which are too small or energies which are too high.
When one proceeds farther along these lines, one gets
into new difficulties and one feels that the basic
ideas necessary to escape from these new difficulties
have not yet been obtained.

The work that has been done since the establishment
of these basic ideas has been important work, but not of
quite the same fundamental standard. People have been
working out the consequences of the early ideas and
examining how far they will go before the difficulties
become serious. The difficulties stem from the fact
that the interaction between elementary particles and
fields is really too violent for a satisfactory theory
to be set up. One has to adopt all sorts of tricks to
make the theory go farther. One has to set up theories
which are more or less patch work and do not have a
fundamental basis.

The present-day situation is that we still have
these fundamental difficulties. It would need someone like a new Heisenberg to find the escape from them.
The experimental people are making steady progress quite
undeterred by theoretical difficulties. They go on ac-
cumulating their evidence and challenging the theoretical
physicists to produce theories which will fit it. The
trouble here is that the experiments are extremely expensive, but in spite of that, experimental work is
stimulated by national rivalry and work is going ahead
in various centers.

My own contributions since these early days have
been of minor importance and I don't think I need mention
any details, except to say that after the establishment
of the existence of the positron I was led to think of a

new particle, the magnetic monopole. There is some very
beautiful mathematics underlying this monopole and it
would make people quite happy if it was found that monopoles do exist in nature, so that this beautiful mathematics has an application. However, I don't have any
fears about this theory if the monopoles are not discovered. If this mathematics is found not to apply to
nature it won't really matter, because this work is quite
isolated and one can abandon it without affecting the
main ideas of the quantum theory.

It is when one is challenging the main ideas that
one has the great excitement and the great fears that
something will go wrong and this sort of excitement has
not recurred since those early days. One might call the
period from 1925 onwards for a few years the Golden Age
of Physics, when our basic ideas were developing very
rapidly and there was plenty of work for everyone to do.
The limitation of the ideas that were established in this
Golden Age have now become clear and we are all hoping that
a new Golden Age will appear, triggered off by some very
drastic new idea and leading once again to a period of
rapid development with great hopes and fears.

JOHN BARDEEN

HISTORY OF SUPERCONDUCTIVITY RESEARCH

John Bardeen

University of Illinois at Urbana-Champaign

Urbana, Illinois 61801

I. INTRODUCTION

When in 1908 Kammerlingh Onnes, working at
Leiden, liquified helium for the first time, he opened
up for investigation the range of very low temperatures
within a few degrees of the absolute zero. Matter
exhibits many remarkable properties at these temperatures, of which superconductivity is perhaps the most
striking. In 1911, Onnes discovered that when cooled
to liquid helium temperature, mercury becomes superconducting, losing all trace of resistance to the flow of
electricity. It was soon found that many metals and
alloys become superconducting below a transition temperature characteristic of the material. A current in
a superconducting ring once started flows indefinitely
with no battery or other source of power. For the discovery of superconductivity, Onnes was awarded the
Nobel Prize in 1913.

Research in superconductivity gradually expanded
from Leiden to many other laboratories throughout the
world and the properties of superconductors have been
extensively investigated. For nearly fifty years,
superconductivity defied explanation and was one of the
outstanding puzzles of physics. We now know that it is
a quantum phenomenon and that superconductors exhibit
quantum effects, ordinarily expected only at the atomic
level, on a macroscopic scale. For this reason, superconductors are called quantum fluids.

Many years after it was first liquified, it was found that liquid helium itself is a quantum fluid. At atmospheric pressure, it remains a liquid down to the absolute zero. Below the λ-point (2.2°K) it is a superfluid and can flow through narrow capillaries with no friction. An illustration of the quantum aspects is that the circulation is quantized to be an integral multiple of h/M, where M is the mass of a helium atom and h is Planck's constant. The corresponding effect in a superconductor is that the magnetic flux threading a superconducting ring is quantized in units of the flux quantum, hc/2e.

The expansion of research during the past decade has been particularly rapid, in part because of a theory that gives a basis for interpretation of experimental data and prediction of new effects, but also because of applications that are beginning to appear. These include superconducting magnets, very sensitive detecting instruments, cavities for particle accelerators, and many others. Serious consideration is being given to the use of superconductors in underground power transmission lines and in electric motors and generators.

How did this expansion from one laboratory in 1908 to the many today come about? How were the key discoveries made and how were the key theoretical concepts developed? In this chapter, we will attempt to trace the main threads of development of research in superconductivity, concentrating on people and places as much as on the physics. Emphasis will be placed on theory rather than experiment, although there has been a close interaction between them.

Since the explanation of superconductivity depends on quantum theory, there was no possibility of doing so prior to 1925-1927 when quantum theory was developed by Heisenberg, Schrödinger, Dirac and others. Quantum concepts were soon applied with great success to account for the properties of normal metals by Bloch and others, but superconductivity remained a puzzle. It was not until 1957 that an adequate quantitative theory was given by L. N. Cooper, J. R. Schrieffer and the author. This theory did not require mathematical methods that were unknown in 1930, but it did require a deep understanding of the physics of superconductors that gradually evolved through a long series of experimental and theoretical researches by many people. A key step was the suggestion by H. Fröhlich, confirmed experimentally

by the isotope effect, both in 1950, that interactions
between electrons and phonons (the quanta of lattice
vibrations) are responsible. Since 1957, the theory
has been further developed by many people and has
stimulated a great deal of new experimental and theoretical research. If the transition temperature is
taken as an empirical parameter, the properties of
superconductors are now as well understood as the
properties of normal metals.

The discovery of new superconducting elements,
compounds and alloys has been done mainly by empirical
methods. The master of this approach has been
B. T. Matthias, who with his associates has found the
vast majority of new superconductors, many with unusual
properties. In 1954, he and others working at the Bell
Telephone Laboratories found that Nb_3Sn is a superconductor with a relatively high transition temperature
of $18^\circ K$. Later, in 1961, J. E. Kunzler, also at the
Bell Laboratories, found that Nb_3Sn remains superconducting in magnetic fields as high as 200,000 gauss,
which made possible the design of high field superconducting magnets.

Another major stimulus to the expansion of research
on superconductivity was the theoretical prediction in
1962 of superfluid tunneling through a thin barrier
separating two superconductors by B. D. Josephson, then
a graduate student at Cambridge University in England.
This prediction, soon verified experimentally, gives a
striking illustration of the macroscopic quantum aspects
of superconductors. It led not only to a deeper understanding of superconductivity, but to the design of
extremely sensitive detecting instruments for voltage,
magnetic fields and other effects.

These are just a few of the highlights of the
expansion of interest in superconductivity. There has
been a close interplay of theoretical and experimental
research, with theoretical predictions often leading
the way to experimental findings and experimental discoveries leading to further understanding. While
scientific interest still remains high, it is expected
that further expansion of the field will be dependent
on the many applications which are beginning to appear.

II.   DISCOVERY OF SUPERCONDUCTIVITY

At the time he made his momentous discovery, Onnes

was professor of physics at the University of Leiden,
where he and his associates concentrated on low temperature physics. With liquid helium available, he
initiated experiments to see how the resistivity of
metals would vary as the absolute zero is approached.
There was some indication that the resistivity of pure
metals should vanish as $T \rightarrow 0^{\circ}K$. He chose to study
mercury because it could be obtained in pure form.   In
his own words, "It could be foretold that the resistance
of a wire of solid mercury would be measurable at the
boiling point of helium but would fall to inappreciable
values at the lowest temperatures which I could reach.
With this beautiful prospect before me there was no
more question of reckoning with difficulties. They
were overcome and the result of the experiments was as
convincing as could be hoped. No doubt was left of the
existence of a new state of mercury in which its resistance has practically vanished....Mercury has passed
into a new state, which on account of its extraordinary
electrical properties may be called the superconductive
state." The transition from the resistive to the
superconducting state occurred abruptly when a critical
temperature of $4.2^{\circ}K$ was reached. It was soon found
that high purity is not required; considerable impurities could be added without affecting the transition
very much.

When other metals were investigated, it was found
that superconductivity is a fairly common phenomenon.
Some of the metallic and nonmetallic elements become
superconducting only at high pressures. A number of
these were discovered only fairly recently. Many alloys
and intermetallic compounds have also been found to be
superconducting. The highest critical temperature so
far attained is $20.9^{\circ}K$ in a ternary compound of niobium,
germanium and aluminum.

To show that the resistance in the superconducting
state really vanishes for all practical purposes, Onnes
showed that persistent currents flow indefinitely in a
ring or solenoid with no battery or other source of
power. In a dramatic demonstration, he transported a
lead ring carrying a supercurrent in a specially prepared cryostat from Holland to England, and showed in a
lecture to the Royal Society that there had been no
decay in the current.

If a magnetic field is applied while in the superconducting state, currents are induced which keep the

flux from changing in the interior of the ring. In the normal state, scattering of electrons would very quickly reduce the current to zero, but in the superconducting state there is no perceptible decrease and the current flows as long as the temperature is maintained below T_c.

Onnes immediately thought of the idea of making superconducting magnets, but these hopes were soon dashed when it was found that superconductivity is destroyed by a critical magnetic field of usually only a few hundred gauss. The critical field, H_c, drops from a maximum at $T = 0$ to zero as $T \rightarrow T_c$. The form of the curve is approximately parabolic, with

$$H_c = H_o(1-t^2) \qquad\qquad (1)$$

where $t = T/T_c$ is the reduced temperature. It is only recently that materials have been discovered that remain superconducting to very high fields.

There is a critical current as well as a critical magnetic field. According to Silsbee's hypothesis, the critical current is that which produces a magnetic field at the surface equal to H_c. This criterion is valid for bulk specimens, but in thin films the critical current density may be independent of the magnetic field. Onnes was able to get current densities as high as 10^5 amps/cm^2; much higher values are now possible.

Onnes retired in the early twenties, when the responsibility for directing the research activities at Leiden were divided between W. H. Keesom and W. J. de Haas, both of whom contributed to further developments in superconductivity as well as working on many other aspects of low temperature physics. Since 1938, with C. J. Gorter as director, the Kamerlingh Onnes Laboratory at Leiden has been concerned mainly with magnetic problems, liquid helium, and other areas of low temperature physics, but has continued work on superconductivity.

III. EXPANSION OF RESEARCH IN LOW TEMPERATURE PHYSICS

For many years, Leiden remained the world center for research in low temperature physics. Through the twenties it was practically the only center. In the late twenties and early thirties, several other groups

got started. At that time Germany was the leading
country in the world in physics research; it was there
that quantum theory was discovered in 1925 and 1926.
There were strong programs in low temperature physics
at Berlin under W. Meissner and F. E. Simon. German
leadership was destroyed when Hitler came to power with
the resulting exodus of many physicists to other countries. Simon, Kurt Mendelssohn and N. Kurti went to
Oxford from Germany to initiate a program in low temperature physics that has been active ever since. The first
helium liquifier in Britain was constructed at Oxford in
1933. Fritz and Heinz London also went to Britain as
refugees; it was there that they developed their phenomenological theory of superconductivity. Fritz London
later went to Duke University and helped establish a
program there. Another refugee who was later to make
very important contributions to superconductivity theory
is H. Fröhlich, now a professor at Liverpool.

Lord Rutherford attracted a young Russian physicist,
Peter Kapitza, to Cambridge University. Kapitza was
interested in high magnetic fields and low temperatures
and became the first director of the Mond Laboratory in
1930. Although Kapitza himself returned to Russia in
1932, the Mond Laboratory has continued to be a leading
center. It is now under the direction of an early
associate of Kapitza, David Shoenberg.

Students from both Oxford and Cambridge have gone
to establish low temperature groups in many parts of
the world. To mention just a few, from Oxford J. G.
Daunt went to Ohio State University and then to Stevens
Institute of Technology, J. L. Olsen went to Zurich,
K. R. Atkins to Pennsylvania and B. S. Chandrashekar to
Case-Western Reserve. Among those who studied at
Cambridge are J. K. Hulm who spent some time at the
University of Chicago and is now at Westinghouse,
W. F. Vinen at Birmingham and B. B. Goodman, formerly
of Grenoble and now with an industrial company in
England. Leaders in superconductivity research present-
ly at Cambridge include A. B. Pippard and B. D.
Josephson.

The first helium liquifier in the Soviet Union was
built by L. W. Shubnikov and associates at the Kharkov
Physico-Technical Institute in 1932. Many leaders in
low temperature research in the USSR were trained at
Kharkov. Among them are N. E. Alekseevski of the
Institute of Physical Problems in Moscow, I. K. Kikoin
of the Kurchatov Atomic Energy Institute and B. G.

Lazarev who now heads the activity at the Physico-Technical Institute. Unfortunately, Shubnikov fell victim to one of Stalin's purges and was not heard from after 1937. He was exonerated twenty years later.

After Kapitza returned to Russia, he became director of the Institute of Physical Problems in Moscow. In the Institute, a strong theoretical group was formed in 1937 under the direction of L. D. Landau. For many years the Institute of Physical Problems has been one of the world's leading centers in both experimental and theoretical aspects of low temperature physics. A few years ago, part of the theoretical group split off to help found the Institute for Theoretical Physics under the direction of I. M. Khalatnikov. Many leaders in theoretical research in the Soviet Union were trained in Landau's group. Those prominent in superconductivity theory include V. L. Ginzburg, A. A. Abrikosov and L. P. Gor'kov. In his early years, Landau spent some time at Kharkov.

The largest center for low temperature research in the Soviet Union, the Physico-Technical Institute of Low Temperatures of the Ukranian Academy of Sciences, was established at Kharkov about 1960 under the direction of B. I. Virkin and B. N. Eselson. There is a total of about 2,000 employees, including several hundred scientists, working exclusively on basic and applied research in low temperature physics. Another important center is at Tbilisi where the Institute for Physics of the Georgian Academy of Sciences is located. The director, E. L. Andronikashvili, is famous for his research on the properties of liquid helium. He worked for a number of years at the Institute for Physical Problems in Moscow before returning to Tbilisi.

One of the earliest centers for low temperature research was established at Toronto under the direction of J. C. McLennan. In the late twenties and early thirties, he and his students did a great deal of work on the properties of superconducting alloys. One of the students, J. F. Allen, went to the Mond Laboratory at Cambridge and later became head of the Physics Department at St. Andrews.

The earliest groups in the USA for work at helium temperatures were established at Berkeley, Yale and M.I.T. Work at Berkeley, under the direction of W. F. Giauque, was concerned mainly with magnetic and calorimetric properties of materials and that at Yale under

C. T. Lane mainly with the properties of helium, so that
little work was done on superconductivity prior to the
second world war. Two students of the many students of
Lane who have later become prominent in low temperature
research are the Fairbank brothers, W. M. now at
Stanford, and H. A., now at Duke.

The real flowering of low temperature research in
the U.S.A., and in fact throughout the world, occurred
after World War II. One of the main reasons was the
development by S. C. Collins at M.I.T. of a helium
liquifier which bypassed the use of liquid hydrogen and
produced liquid helium directly. This machine, put in
production by the A. D. Little Company, allowed any
laboratory to get into the field at modest cost.

Expansion of low temperature research in the U.S.A.
was actively promoted by the Office of Naval Research,
which was the first agency to support research in the
universities after the war. Recognizing that this was
an important and growing field of research, they helped
a number of universities to do research in the low
temperature field. A key person in this activity was
L. M. McKenzie. The ONR helped sponsor a series of
semiannual conferences at which information on scientific results and techniques could be exchanged. In
the late forties, nearly all of the research in the
U.S.A. in low temperature physics was supported by the
ONR. Some of the important groups in addition to those
at Yale, M.I.T. and California were at the National
Bureau of Standards, Columbia, Rutgers, Duke, Ohio
State and Illinois. Now most of the world's leading
physics laboratories have facilities for low temperature
research. The 300th Collins machine was installed at
St. Andrews in 1965.

In the early fifties, a number of industrial
laboratories, including General Electric, Westinghouse,
IBM, RCA, and the Bell Telephone Laboratories initiated
programs in superconductivity and other aspects of low
temperature research. With appropriations from the ONR
and other Defense Department agencies leveling off, the
National Science Foundation and later the Atomic Energy
Commission became important sources of support for
university research.

My own research at Illinois has been supported
since 1952 by the U. S. Army Research Office at Durham,
North Carolina. This office was just getting started
at about the time I arrived at Illinois in the fall of

1951, so that funds were available.

IV.   RESEARCH IN THE THIRTIES

With research groups at Leiden, Oxford, Cambridge, Kharkov, Moscow, Toronto and Berlin, the thirties was an active period for superconductivity. It was one of the main subjects for discussion at international physics conferences. Many of the world's leading theorists attempted without success to find an explanation for this puzzling phenomenon. In the absence of a fundamental explanation, attempts were made to find phenomenological descriptions of the thermal and electromagnetic properties.

The most important discovery on the experimental side is that of Meissner and Oschsenfeld who showed in 1933 that a superconductor is a perfect diamagnet and excludes a magnetic field. Even if the specimen is cooled into the superconducting state in the presence of a magnetic field, the flux is expelled, so that the state with the flux excluded is the unique thermodynamically stable state. This work gave justification for application of thermodynamics to derive relations between critical fields and specific heats, as was done with great success by Keesom, A. J. Rutgers and particularly by Gorter.

If the flux is excluded, there is an increase in magnetic energy of $H^2/8\pi$ per unit volume. If this increase more than makes up for the free energy difference between normal and superconducting states, a transition to the normal state will occur. Thus the critical field, H_c, is given by

$$G_n - G_s(0) = H_c^2/8\pi , \tag{1}$$

where G_n and $G_s(0)$ are the free energies of the normal and superconducting states in the absence of a magnetic field.

Keesom, van den Ende and Kok showed about the same time that in zero magnetic field, the normal-superconducting transition is of second order with a jump in specific heat but no latent heat at the transition temperature. Such a behavior is consistent with a critical field curve of the sort indicated in Eq. (1).

To account for the thermal and transport properties,

Gorter and H. B. G. Casimir proposed a two-fluid model
in which the particle and current densities are
expressed as the sum of normal and superfluid components:

$$n = n_n + n_s \qquad (2)$$

$$J = n_s e v_s + n_n e v_n. \qquad (3)$$

The component, n_s, which represents the superfluid
condensate, is equal to the total electron density, n,
at $T = 0°K$, but decreases with increasing temperature
and goes to zero at T_c. The normal component, subject
to the usual dissipation, comes from electrons excited
out of the condensate.

Not long after, in 1935, the London brothers
extended the two-fluid model to give their famous phenomenological description of the electromagnetic properties. Consistent with the Meissner effect, they
proposed that the superfluid current density, $J_s = n_s e v_s$,
is determined uniquely by the magnetic field, H,
according to the equation

$$(4\pi\lambda^2/c)\,\mathrm{curl}\ J_s = H \qquad (4)$$

and that it is accelerated by an electric field, E,

$$(4\pi\lambda^2/c^2)\,dJ_s/dt = E, \qquad (5)$$

where $\lambda^2 = mc^2/4\pi n_s e^2$ is the square of the penetration
depth. Although the London equations are now known not
to be exact, they have been used ever since to give a
qualitative and in many cases semiquantitative description of superconductivity.

According to these equations, the magnetic field
drops exponentially to zero with distance, x, from the
surface of a bulk specimen:

$$H = H_o e^{-x/\lambda}. \qquad (6)$$

The penetration region, of thickness of order λ, is
where currents are flowing to give a field to counteract
the external field in the interior. Typically, λ is of

the order of a few hundred Angstroms ($\sim 5 \times 10^{-6}$ cm).

    At a discussion of superconductivity held at the
Royal Society in May 1935, Fritz London suggested that
the explanation of superconductivity depends essentially
on quantum theory and indicated how the phenomenological
equations might follow from a quantum approach. Following the Meissner effect, he took the point of view that
the diamagnetic aspects are basic and suggested that the
entire superconductor behaves as a "single big diamagnetic atom." He supposed that "the electrons be coupled
by some sort of interaction in such a way that the
lowest state may be separated by a finite interval from
the excited ones." This may be the earliest suggestion
of an energy gap for quasiparticle excitations from the
superconducting ground state, a feature of the present
microscopic theory. It has turned out that Fritz London
was correct in his conjecture that superconductivity is
a quantum phenomena, and he has been very influential in
pointing the way towards a microscopic theory.

    Shortly after these developments in Western Europe,
Landau worked out a theory of the intermediate state as
a series of alternating lamina of normal and superconducting regions. The flux is carried in the normal
regions where the field is approximately equal to H_c.
In a body of any shape, except a long cylinder in a
magnetic field parallel to the axis, the intermediate
state occurs over a range of magnetic fields prior to
a complete destruction of superconductivity. For
example, for a sphere in a uniform field, the field at
the equator is 3/2 the applied field if the flux is
excluded from the interior. Thus when $H_a = \frac{2}{3}H_c$, the
field reaches the critical value at the equator, but is
less than H_c at other points. For $\frac{2}{3}H_c < H_a < H_c$, the
sphere enters the intermediate state. With little in
the way of experimental evidence to go on, Landau
developed his picture of the structure of the intermediate state. His theory was confirmed shortly after
the war in a beautiful series of experiments done in
Moscow.

    At about the same time, in 1936 and 1937, Shubnikov
discovered a new class of superconductors, which he
called type II. In these, flux starts to penetrate at
a lower critical field H_{c_1}, which may be well below the
thermodynamic critical field, H_c, but the metal remains

superconducting to an upper critical field, $H_{c_2} > H_c$.
Before he had an opportunity to complete these studies,
Shubnikov was purged. His work did not receive the
attention that it should have until twenty years later.

Further important work in the pre-war years was
concerned with the electrodynamics of superconductors.
The first estimates of penetration depths were made by
M. von Laue from measurements of R. B. Pontius on the
critical fields of thin lead wires. Shoenberg made a
series of measurements on the magnetization curves of
colloidal suspensions of mercury with particle size
between 10^{-6} and 10^{-5} cm. Because of a spread in particle size, the measurements were difficult to interpret,
but they were consistent with an increase of penetration
depth with temperature following the law

$$\left(\frac{\lambda(t)}{\lambda(0)}\right)^2 = \frac{1}{(1-t^4)} , \tag{7}$$

where $t = T/T_c$ is the reduced temperature. This law
follows from the London theory if it is presumed that
n_s varies with temperature in the way suggested by the
Casimir-Gorter two-fluid model. In the years 1938-1940,
various groups studied the magnetic properties of thin
films with thicknesses comparable to the penetration
depth.

The work we have discussed so far involves static
magnetic fields. The first high frequency measurements
were carried out by H. London in 1940. He measured the
surface resistance, R_s, at microwave frequencies and
found that R_s is not discontinuous at T_c, but drops
rapidly from the value in the normal state at $T = T_c$
and approaches zero gradually as $T \to 0^\circ$K. London
reasoned that the high frequency electric fields in the
penetration region should accelerate the normal electrons and give a loss. Since according to the two-
fluid model, n_n should vary continuously across T_c, so
should R_s. At that time microwave techniques were in
their infancy, so that the methods used were primitive
according to present standards, but he found at least
qualitatively the expected behavior. Measurements of
surface impedance at high frequencies became an important method for studying superconductors in the post-
war years.

Most of the developments of the period we have been

discussing were included in a small but excellent book
by Shoenberg, published in 1938. Of Russian extraction,
he was very familiar with Russian work and helped make
it better known in the West. It was his book more than
any other source that helped introduce me to the field
of superconductivity.

V. POST-WAR RESEARCH (1946-1957)

Although there was little work on superconductivity
done during the war, a vigorous program was resumed at
Cambridge, Moscow, and other centers in the immediate
post-war years. Microwave techniques had been greatly
advanced by research done for radar and were applied,
particularly by A. B. Pippard and co-workers at
Cambridge, to study the electrodynamics of superconductors.

With separated isotopes available from the atomic
energy program, studies were initiated at Rutgers
University and at the National Bureau of Standards to
see whether or not the transition temperature depends
on the isotopic mass. This led, surprisingly, to a
positive result and showed that electron-phonon interactions must play a predominant role in superconductivity. Independently, on theoretical grounds, this
was suggested by H. Fröhlich. In Russia, experiments
of Shalnikov and co-workers confirmed Landau's con-
jectures about the intermediate state. In 1950,
Ginzburg and Landau extended the London phenomenological
theory with the introduction of a complex order parameter with amplitude and phase. This theory was applied
by Abrikosov to account for the experimental results of
Shubnikov on type II superconductors. Several laboratories in the U.S.A. became involved in superconductivity research and made important contributions to our
understanding. In this section we shall discuss these
developments that formed the groundwork for the microscopic theory.

Another important event in 1950 was the publication
of Vol. I of F. London's book on Superfluids, which
after 20 years is still the most complete account of the
London phenomenological theory of the electromagnetic
properties of superconductors. He acknowledges the
support of the U. S. Office of Naval Research, which
allowed him to visit English and Dutch laboratories and
get the benefit of the latest advances in those coun-

tries. In the book, he explained his ideas on what sort
of a microscopic theory would be required to explain
superconductivity, making use of quantum concepts. His
ideas are remarkably close to the present theory. He
states that "the long range order of the average momentum is to be considered one of the fundamental properties of the superconducting state." This would result
from "the wide extension in space of the wave functions
representing the same momentum distribution throughout
the whole metal in the presence as well as the absence
of a magnetic field." Superconductivity is "a quantum
structure on a macroscopic scale" which requires "a
kind of solidification or condensation of the average
momentum distribution." His ideas as expressed in his
book and expounded in conversations and lectures had a
strong influence on subsequent developments.

After the second world war at Cambridge, under the
leadership of Shoenberg and Pippard, a number of experiments were carried out to measure penetration depths
and the high frequency behavior of superconductors.
While some experiments involved thin films and small
specimens, others were done on bulk material. Shortly
before the war, Casimir had proposed that one could
measure the change in penetration depth with temperature
by measuring the temperature variation of the inductance
of a coil closely wrapped around the specimen in the
form of a circular cylinder. With increasing penetration of field into the specimen with increasing temperature, the inductance should increase as T_c is approached. Laurent and Shoenberg in 1947 applied this method
and found a temperature variation consistent with (7).
Making use of this relation, they were able to estimate
$\lambda = 5 \times 10^{-6}$ cm for tin.

In an extensive series of measurements, Pippard extended H. London's microwave method with much improved
techniques to measure both surface resistance and surface reactance as functions of frequency. Changes in
reactance are related to changes in penetration depths,
so that he was able to determine the temperature variation of λ. The studies include effects on penetration
depths of a static magnetic field and of alloying to
change the mean free path, of the conduction electrons.
He found that a decrease in m.f.p. gave a marked increase in λ even though the density of electrons was
hardly affected at all, a result inconsistent with
the London theory. These and other results suggested
to him a nonlocal modification of the London equations

in which a coherence distance, ξ_0, is introduced. The
current density is not proportional to the vector potential, A, but is given by an integral of A over a region
of $\sim\xi_0$ surrounding the point in question. It is the
Pippard nonlocal version that is confirmed by present
microscopic theory.

Other experiments provided evidence for an energy
gap in the quasiparticle excitation spectrum of superconductor. In a normal metal there is a continuous
distribution of levels. At T = 0°K, those below the
Fermi level, E_F, are occupied, those above unoccupied.
Excitations correspond to exciting particles into states
above E_F, leaving unoccupied states or holes behind.
Such a picture leads to a specific heat of the electrons
proportional to T at low temperatures, as is observed
experimentally in normal metals. The picture is qualitatively the same when interactions between electrons
are taken into account.

In an energy gap model, the electrons are presumed
to go into some sort of condensed state such that a
finite energy, the energy gap, E_g, is required to excite
a particle from the sea. If such a picture is valid,
the number of excitations at low temperatures should
vary as $\exp[-E_g/2k_BT]$. The first convincing evidence
for such a dependence came in 1951 from B. B. Goodman at
Cambridge on the thermal conductivity of tin. Not long
after, an exponential decrease in specific heat as
T → 0°K was found by Brown, Zemansky and Boorse for
niobium at Columbia and by Corak and co-workers for
vanadium at Westinghouse.

If there is an energy gap, it should show up in the
frequency dependence of the absorption as T → 0°K. If
$\hbar\omega < E_g$ there should be no absorption at T = 0°K because
there are no quasiparticle excitations. Experiments at
microwave frequencies at very low temperatures had indicated that R_s → 0 as T → 0°K and those done at optical
and infrared frequencies had indicated no difference
in reflectivity and thus in absorption from the metal in
the normal state. If E_g is of the order of a few times
k_BT_c, the critical frequency at which absorption should
set in is in the very far infrared, the most difficult
part of the spectrum to work in. Working at the University of California at Berkeley, M. Tinkham and his
student R. E. Glover, III, developed techniques for use
in this region and measured the transmission through
thin lead films as a function of frequency. They found
that absorption sets in at quantum energies between

$3.5 \ k_B T_c$ and $4.0 \ k_B T_c$. They also measured the reactive
component and found that the results could be interpreted in terms of the Pippard nonlocal theory, and were
able to estimate a value for the coherence distance, ξ_o,
in fair agreement with earlier estimates made by Pippard
and Faber at Cambridge. These experiments, which gave
convincing evidence for an energy gap, were done in 1956
and 1957, about the same time as the development of the
microscopic theory. These and later experiments provided a very good confirmation of the theory.

VI.   ELECTRON-PHONON INTERACTIONS

Experiments on the isotope effect were very important because they indicated that the motion of the ions,
and thus the lattice vibrations are involved. The
experiments were done independently by a group at Rutgers
University under the leadership of B. Serin and by
E. Maxwell working at the National Bureau of Standards.
Separated isotopes had become available as a result of
the atomic energy program. It was decided to see whether
or not the transition temperature depends on isotopic
mass. There was no prior indication that a positive
result was to be expected. The experiments were done
as a shot in the dark, with the probabilities indicating
a null effect. Lattice vibrations, or their quanta, the
phonons, had not been considered as a possible mechanism.

The isotope effect was first announced by the two
groups at the ONR sponsored meeting on low temperature
physics held at Georgia Tech in March, 1950, but was not
publicized outside of the low temperature community. I
first learned about it in a telephone call from Serin in
May. He reported that he and his coworkers had found
that T_c varies inversely with the square root of the
isotopic mass. When I heard about this result, I thought
that electron-phonon interactions must be involved and
attempted to construct a theory on this basis.

My first attempt at a theory of superconductivity
had been made in the late thirties and was strongly
influenced by London's ideas. I thought that small
complex lattice distortions might create energy gaps at
the Fermi surface in such a way as to lower the energy
of electrons in states just below the Fermi level.
Electrons in such states would have a very small effective mass, large orbits and a large diamagnetic effect.
The attempt was not successful, and only an abstract of

the work was published, but it did have some attractive features. The isotope effect suggested that the energy gap might arise from dynamic interactions with the lattice vibrations, or phonons, rather than from static lattice distortions.

In June, about a week after I had submitted a letter to the Editor of the Physical Review outlining these ideas, Fröhlich visited the Bell Telephone Laboratories, where I was working at the time. I was surprised to learn that when he had been visiting at Purdue University during the spring term he had worked on a theory of superconductivity based on electron-phonon interactions and had recently submitted his paper for publication. Using a fieldtheoretic approach, he showed that the electron-phonon interaction leads to an effective attractive interaction between electrons near the Fermi surface. Such an interaction should affect the energy of the electrons, and his theory was based on the self-energy of the electrons in the phonon field. He was greatly encouraged when he learned, just about the time he was ready to send his manuscript to The Physical Review, about the isotope effect, which gave a strong experimental confirmation of his approach. Although there were mathematical difficulties in both his method and mine, we were both convinced that at last we were on the way to an explanation of superconductivity.

This was indeed the case, but it turned out that the difficulties were basic and not easy to overcome. I wrote an article for the Reviews of Modern Physics about a year later (1951) which outlined the serious problems involved and the attempts which had been made to overcome them. We now know that the trouble with these early theories of Fröhlich and myself is that they were based on self-energy rather than a true interaction between electrons. Another difficulty was the use of perturbation theory. Later, it was shown by M. R. Schafroth that a theory based on treating the electron-phonon interaction by perturbation theory could not account for the Meissner effect even though the expansion is carried to arbitrarily high order. In 1956, A. B. Migdal showed that the phonon-induced self-energy is included in the normal state and does not change the character of the Fermi surface.

With the microscopic theory running into a stone wall, I turned my attention to other problems. Interest was renewed in about 1955, when I was invited to write

a review of the theory of superconductivity for the
Handbuch der Physik. Although at the time there was no
way to derive it from microscopic theory, there had
been increasing experimental evidence for the energy
gap model and this was the unifying theme of the article.
Fröhlich's derivation of an effective attractive interaction between electrons from exchange of virtual phonons had been extended by David Pines and me to include
Coulomb interactions. It was suggested that one should
take the complete interaction, not just the diagonal
self-energy terms, and use it as the basis for the
theory.

VII. MICROSCOPIC THEORY

It was becoming evident that fieldtheoretic methods,
developed mainly for high energy physics, would be a
valuable tool for solving the many-body problems of a
Fermi gas with attractive interactions between the
particles. In the spring of 1955, I called C. N. Yang,
who was then at the Institute for Advanced Study at
Princeton, and asked if he knew of someone versed in
field theory who might be willing to work on superconductivity. He suggested Leon N. Cooper, who had
been spending the year there as a postdoctoral fellow
after getting his Ph.D. from Columbia in nuclear theory.
Cooper arrived at Illinois in the fall of 1955 for a
fresh attack on the microscopic theory. J. R.
Schrieffer, a brilliant graduate student who had done
his undergraduate work at MIT, was about ready to start
a thesis problem and wanted to join in. He preferred
to gamble on this difficult problem rather than pick a
more straight-forward topic for his thesis. Cooper and
I shared an office and Schrieffer was a frequent visitor,
so that we had a close collaboration.

Cooper made a major step forward when he showed
that in the presence of an attractive interaction two
electrons excited out of the Fermi sea will form a
bound state no matter how weak the attraction. The
Fermi sea would then be unstable against the formation
of such bound pairs. It had earlier been suggested by
V. L. Ginzburg and by M. R. Schafroth that superconducting properties would result if electrons combine
into even groupings so that the resulting aggregates
would obey Bose statistics.

However, it was not possible to use this approach
immediately to construct a theory of superconductivity.

If the binding energy of a pair is of the order $k_B T_c$, the size of the pair wave function is of the order of 10^{-4} cm. But if all electrons within $\sim k_B T_c$ of the Fermi surface take part in pairing, the average spacing between pairs would be only about 10^{-6} cm, a distance much smaller than the size of the pair. Thus a true many-body approach was required to find the ground state wave function of a superconductor. Schrieffer played a major role in finding a variational form for the wave function. We all collaborated closely in working out the quasiparticle excitation spectrum and in applying the theory to a number of problems.

According to the theory, the ground state can be regarded as a linear combination of normal configurations in which the normal quasiparticle states are occupied in pairs $(k_1 \uparrow, k_2 \downarrow)$ of opposite spin and the same momentum $k_1 + k_2 = q$, common to all pairs. If in a given confirguration one of the pair is occupied, the other is also. It is the common momentum of the pairs that defines a state of macroscopic occupation and gives the long range order of the momentum distribution, suggested by London as a characteristic feature of a superconductor. The pairing makes maximum use of the available phase space to give a coherent low-energy state. Values of q different from zero describe current flow in the ground state, with $2p_s = 2mv_s = \hbar q$.

A preliminary note was submitted in February, 1957, and published in the April 1, 1957 issue of the _Physical Review_. This letter included an expression for the ground state energy difference between normal and superconducting states and for the energy gap at $T = 0°K$. During the spring, Cooper, Schrieffer and I were busy extending the theory to obtain the excitation spectrum and applying it to derive various thermal and transport properties. We worked out an expression for the specific heat and showed that the transition at T_c is of second order with a jump in specific heat but not latent heat. Many of the results were given by Cooper in a talk at a special session of the March, 1957, meeting of the American Physical Society.

A great simplification is that the quasiparticle excitation spectrum of a superconductor is in one-to-one correspondence with that of the normal state. If ε_k is the normal state quasiparticle energy relative to the Fermi surface, the energy of the corresponding excitation in the superconducting state is $E = \sqrt{(\varepsilon_k^2 + \Delta^2)}$, where Δ is the gap parameter. Since Δ is a measure of the pairing, it is also called the pair potential. With increasing temperature, Δ decreases and goes to zero at $T = T_c$. Quasiparticle excitations are created in pairs from the ground state; the minimum energy required to create a pair is the energy gap $E_g = 2\Delta$. The temperature variation of Δ predicted by the theory has been confirmed strikingly by experiment.

A calculation of the response to static fields showed that there is a Meissner effect, and we derived an expression for the penetration depth and its temperature variation in good agreement with experiment. We found an expression for the current density similar to that suggested by Pippard on phenomenological grounds and derived an expression for the Pippard coherence distance, also in good agreement with experiment. Because of the energy gap, there is no absorption of energy in a.c. fields until the frequency becomes greater than the gap, $\hbar\omega > 2\Delta$, beyond which absorption sets in rapidly. Our calculations were in excellent agreement with the experiments of Glover and Tinkham which were just being completed.

While we were working out consequences of the theory in the spring of 1957, Hebel and Slichter, also working at Illinois, were making the first measurements of nuclear spin relaxation times in the superconducting state. A nuclear spin relaxes by interchanging spin with a conduction electron, so that the inverse relaxation time is a measure of the electron-nuclear scattering interaction. It was found surprisingly that the relaxation rate increases as the temperature drops below T_c, indicating a larger interaction in the superconducting than in the normal state. Specific heats and other measurements indicate that the number of excited quasiparticles drops rapidly below T_c, so that it was at first difficult to understand the larger interaction. When the microscopic theory was worked out, it was found that there are important coherence effects in the scattering of quasiparticles in the superconducting state. Because of the pairing, scattering of $\underset{\sim}{k} \uparrow$ is

coherent with scattering of $-k \downarrow$. Depending on the
nature of the interaction, the two contributions can add
constructively or destructively. Ordinary scattering,
which we called Case I, gives destructive interference,
but for spin-flip scattering, called Case II, the two
contributions add constructively to give an abnormally
large relaxation rate.

An example of ordinary scattering is that by longitudinal acoustic waves which results in attenuation
of the waves. At an international conference on low
temperature physics in Madison, Wisconsin in late
August, 1957, R. W. Morse and H. V. Bohm and also
H. E. Bömmel reported on measurements of longitudinal
ultrasonic attenuation in superconducting aluminum.
Instead of rising, the attenuation drops rapidly below
T_c. While there, I discussed the problem with Morse and
others and worked out a theory of the effect. Morse,
then at Brown University is now President of Case-
Western Reserve. The excellent agreement of the theory
for both of the seemingly contradictory experiments on
nuclear spin relaxation and on ultrasonic attenuation
gave an early confirmation of the pairing hypothesis.

For simplicity, our calculation of thermal, electromagnetic and other properties of a superconductor were
based on a simplified model which involves three parameters: the normal density of states in energy, the
velocity of the electron at the Fermi surface and an
interaction constant which gives a measure of the phonon-induced attraction between electrons. Only the
latter involves the superconducting state and can be
related to the transition temperature. The parameters
can be combined to give two lengths, the London penetration depth, λ, and the Pippard coherence distance,
ξ_0. The ratio of these, $\kappa = \lambda/\xi_0$ is an important parameter for a superconductor and determines whether it is
type I or type II.

There is a rough law of corresponding states for
superconductors. When plotted in terms of reduced
coordinates, of which one is the reduced temperature,
$t = T/T_c$, the properties of most superconductors are
pretty much alike. It was found that our model fits the
law of corresponding states about as well as real metals.
In fact we were continually amazed at the excellent
agreement between theory and experiment. If there was
a discrepancy, it was usually found on rechecking that
an error had been made in the calculations. All of the
hitherto puzzling features of superconductors fitted

neatly together like the pieces of a jigsaw puzzle.

VIII.   RECEPTION OF THE THEORY

Because the theory had the long range coherence of
momentum and other basic features that London and others
had thought a microscopic theory should have, and because
we were able to get such good quantitative agreement for
a wide range of phenomena, we were convinced that we had
an adequate explanation of superconductivity. Accordingly, we were surprised at the skepticism with which
the theory was received in some quarters. Some, such
as P. W. Anderson and D. Pines, immediately accepted
the theory and made important advances. Many others,
however, expressed doubts that a valid explanation of
superconductivity had been found. Some dismissed the
agreement between theory and experiment as something
that would follow from any reasonable model that gave
an energy gap.

Much of the criticism was directed against our
calculation of the Meissner effect, because it was not
carried out in a manifestly gauge invariant manner.
Our method required a transverse gauge, that used by
London, because response to longitudinal fields would
have required consideration of longitudinal collective
modes which we wished to avoid. Similar objections had
been made and answered in regard to the calculation of
the Meissner effect from an energy gap model in my
Handbuch article. Anderson, in fact, soon introduced
collective coordinates into the microscopic theory and
showed how the calculation could be made in a general
gauge. More complete gauge invariant calculations were
given later by Y. Nambu, G. Rickayzen and others and it
is now generally conceded that our calculation of the
magnetic response is correct.

Doubts were also expressed about persistent currents, although we did give a rather brief explanation
in our paper. The key is the common momentum of the
pairs, which gives a state of macroscopic occupation
and a long range order in momentum, as envisaged by
London. This state of macroscopic occupation is not
changed by scattering of individual electrons, which
can only give fluctuations about the supercurrent flow.
A major disturbance, such as passage of a quantized
vortex line across the specimen, is required to change
the momentum of the condensate, p_s. An energy gap is

not required for persistent current flow.

Most skeptical were those who had worked long and hard on superconductivity theory themselves and had their own ideas on what the theory should be like. In the summer of 1957 there appeared a paper by Blatt, Butler and Schafroth that expanded on Schafroth's earlier ideas on superconductivity as arising from an Einstein-Bose condensation of pairs. They used what they called a quasi-chemical approach in which the ground state wave function is expressed as an antisymmetrisized product of pair functions. In July, I received a letter from F. Dyson, in which he referred to their work and showed that if projected to give a state with exactly N pairs, our variational function could be expressed in just their form. Thus he felt that the theories are identical in principle.

The following is taken from my reply to Dyson dated July 23, 1957:

"As I see it, there are three rather than two concepts which Schafroth, Blatt and Butler have introduced into the theory of superconductivity: (a) a correlation length, which we both feel is incorrect, (b) the idea that a perfect charged boson gas has superconducting properties, and that actual superconductors resemble such a gas, the bosons being bound or quasi-bound electron pairs, and (c) the very elegant formalism called the quasi-chemical approach. I have high regard for the latter, but as far as I am aware, it is so far only a suggested program, with no actual calculations for a model with superconducting properties. I have not seen the article in Helvitica Physica Acta (our copy has apparently not arrived); it is possible this goes beyond the preprint which you referred to. When Blatt and Butler were in this country last fall, they gave various talks in which the formalism was presented, and then they presented qualitative arguments based on (b) above. It is this latter to which we object and is the reason we decided to put our footnote. As I told Blatt, I thought his mathematics was fine, but I didn't believe the talk

that went with it. Since they talk about
pairs, and we talk about pairs in a different sense, we thought confusion would
arise unless we put a note in to help
clarify the situation, but apparently we
have not done so.

"We believe that there is no relation
between actual superconductors and the
superconducting properties of a perfect
Bose-Einstein gas. The key point in our
theory is that the virtual pairs all
have the same net momentum. The reason
is not Bose-Einstein statistics, but
comes from the exclusion principle; one
can make the best use of the available
phase space in this way. A critical
point is that our excitation spectrum
bears no relation to that of a Bose-gas;
we have an energy gap with a very high
density of states just above the gap, a
Bose gas has a density of states varying
as $\sqrt{E}$. Our wave functions indicate that
there is no localized pairing of electrons
into "pseudo-molecules." There is, of
course, some correlation between positions
of electrons of opposite spin, but the
correlation occurs over such large distances that the electrons can not be
grouped into localized pairs (as we
mentioned in our manuscript, there are
of the order of 10^6 interacting electrons
within a coherence volume)."

At present those who like to use the analogy with
an Einstein-Bose condensation of pairs say that the
concept applies only to the condensate, not to the
excitation spectrum. While one can of course use any
analogy that helps understanding, provided that the
limitations are realized, I have always felt that it is
somewhat misleading to call superconductivity an
Einstein-Bose condensation of pairs. The form of the
wave function, with the all-important common momentum
for the paired states, is determined by energetic rather
than purely statistical considerations.

The energy gap is interpreted as the energy required to break up a pair. In the pairing theory,

$$\Delta \sim \hbar\omega_c \exp [-1/N(o)V], \tag{8}$$

where ω_c represents the cut-off of the phonon interaction and is of the order of the Debye energy, $N(o)$ is
the density of states in energy at the Fermi surface and
V is the average of the attractive interaction. The
energy of a single pair in the Cooper problem is of
order

$$E \sim \hbar\omega_c \, \exp \, [-2/N(o)V], \tag{9}$$

a much smaller quantity. To get the correct energy, it
is necessary to solve the many-body problem, and it is
an oversimplification to think of the phenomenon simply
as an Einstein-Bose condensation.

Further, the pairing refers to normal state quasiparticles which include all of the correlation energy
present in the normal state. The correct wave functions
for a superconductor which takes into account the correlations present in the normal metal could not be written
as an antisymmetrized product of pair functions.

Blatt and Matsubara were eventually able to derive
superconducting properties using the quasi-chemical
approach, but it turned out to be quite difficult. Because we did not have boson-like excitations, Blatt and
coworkers rejected our version of the microscopic theory
for a considerable period of time. Reservations were
still being expressed at a meeting on superconductivity
held at Cambridge University in the summer of 1959.

I spent a month at Berkeley in the summer of 1957
where I had many interesting discussions of superconductivity theory with Dyson and Tinkham and others.
Later in the summer I had an opportunity to present the
theory to an international audience at the meeting at
Madison, Wisconsin in August. Neither Cooper nor
Schrieffer could attend, so that after it was over I
wrote to them to give my reactions. Since the letter
expresses something of the flavor of the discussions
that took place, I quote from my letter to Schrieffer
dated September 19:

"The Madison meeting was quite interesting and well attended. There was considerable interest in the superconductivity
theory. While very few had had a chance
to study the theory carefully, the reaction
was generally favorable. Experimentalists
were particularly enthusiastic. Objections
were raised mainly by those with precon-

ceived notions of what the theory should
be like.

Schafroth criticized our derivation
of the electrodynamic properties because
we have a momentum dependent cutoff in
our interaction and we do not take this
momentum dependence into account in our
expression for the current density.
Further, our theory is not obviously
gauge invariant, and it is not easy to
make it so (we would have to include
states with pairing $(k_1\uparrow, k_2\downarrow)$ where
$k_1 + k_2 = q\neq0$). He went further and
stated that there is no way to define
a current with our interaction, but I
don't believe that this is true. Since
in the weak coupling limit we get a
negligible contribution in the neighborhood of the cut-off, I believe his criticism is not valid, but perhaps it would
be worth taking a closer look.

Some of the English (Kuper and others)
took the viewpoint that the check of our
theory did not mean much because almost
any energy gap model would lead to equivalent results. The check found applies
to the energy gap model, not to our specific form of the theory. A very good
answer came out of discussions with Morse
and Bömmel in regard to ultrasonic absorption, for which coherence of the
matrix elements follows case I rather than
case II.

I worked out the expression for
energy absorption for case I, and got
the surprisingly simple result that for
$h\nu<<kT_c$, $\alpha_s/\alpha_n = 2f(\Delta)$, where f is the
ordinary Fermi-Dirac function and Δ is
the energy gap parameter. This should
apply to ultrasonic absorption when the
m.f.p. of the electrons is large compared with wave length of the sound
wave, so that the process is simply
emission and absorption of phonons by
the electrons. After I returned here,
I received a copy of a letter which
Morse and Böhm (of Brown University)

prepared for the Phys. Rev. They used
the theoretical expression to determine
$\Delta(T)$ from their data on ultrasonic
absorption in an indium specimen, and
found excellent agreement with our
theory, as indicated on the enclosed
copies of figures from the letter.
It is perhaps significant that the
estimates from measurements were made
before they received the theoretical
curve.

Thus for $h\nu \ll kT_c$, there is an
abrupt drop in absorption when coherence follows case I (ultrasonic waves)
and an increase just below T_c for case
II (Hebel and Slichter, for NMR and
Tinkham et al., for electromagnetic
waves). This seemed sufficiently important to include in our paper, so I
have added three pages to appear at
the end of Sect. IV."

As we have mentioned, the difficulties concerning
the Meissner effect have since been resolved. A more
complete calculation of the electromagnetic response was
carried out in collaboration with a graduate student,
D. C. Mattis, and submitted for publication early in
1958. We derived an expression valid for all frequencies and temperatures. To include effects of
scattering, we paired normal quasiparticle states in
wave functions appropriate to the scattering centers;
that is scattering was included in zero order. It was
found that a m.f.p. for scattering enters the expression
for the current density in just the way that had been
suggested by Pippard on phenomenological grounds.
Independently, Anderson took impurity scattering effects
into account in his theory of dirty superconductors,
and showed that nonmagnetic scattering has very little
effect on the transition temperature.

IX. RESEARCH IN RUSSIA

A preprint describing the theory and applications
was issued in June, 1957, and had wide circulation in
the West, but we were unable to send copies to Russia.
Our paper in the Physical Review did not appear in
print until November. The only information available
in the USSR was the Letter to the Editor which had

appeared in April. This Letter had attracted much
interest, particularly by Landau and his group at the
Institute for Physical Problems in Moscow and by N. N.
Bogoliubov and coworkers. Bogoliubov, a leading Soviet
mathematical physicist, was able to derive the ground
state and excitation spectrum by a different method than
the one we used, one that used a quasiparticle transformation mathematically similar to one he had given
several years earlier for Bose systems. He and his
coworkers, D. N. Zubarev and I. A. Tserkovnikov, derived
the thermal properties using this transformation.
Bogoliubov also was able to start from a Hamiltonian
which included electron-phonon interactions rather than
an effective phonon-induced interaction and to take the
repulsive Coulomb interactions into account in a more
satisfactory way. It was shown that Coulomb interactions
can lead to an exponent, α, in the isotope effect,
$T_c \sim M^{-\alpha}$, different from the value 1/2 typical of phonons
alone. In November 1957, shortly after they had submitted their first papers for publication, a copy of the
preprint was seen and it was found that where we over-
lapped, their results were identical to ours.

During this time, the Landau group repeated some of
the calculations we had made in regard to the electrodynamic properties. A Soviet scientist who had been to
a summer school at Varenna, Italy, had picked up a copy
of our preprint and brought it back with him, but filed
it away without reading it. When he heard about some
of the results of the Bogoliubov and Landau groups, he
looked up the preprint and then it was found that most
of the results had already been obtained, although by
different methods. Meanwhile, J. G. Valatin, at
Birmingham, independently derived the quasiparticle
transformation of Bogoliubov.

A major contribution of the Landau group is the
introduction of Green's function methods to superconductivity theory, mainly by L. P. Gor'kov, a young
theorist. This formulation has been exceedingly fruitful to the subsequent development of the theory. It
allows one to introduce lifetime effects and to treat
space variations of the order parameter. Similar methods
were developed by P. Martin and J. Schwinger and introduced to superconductivity theory by L. P. Kadanoff and
Martin. The origins of the method go back to Matsubara
who treated inverse temperature as an imaginary time.
In an important application, Gor'kov showed that near
T_c the microscopic theory yields the Ginzburg-Landau
phenomenological theory with the gap parameter $\Delta(r)$

playing the role of an effective wave function.

Another major application of Green's function methods is the theory of Abrikosov and Gor'kov on effects of magnetic impurities which can cause spin-flip scattering. While scattering by ordinary nonmagnetic impurities has relatively little effect on the transition temperature, additions of small amounts of magnetic impurity can cause T_c to decrease and eventually to destroy superconductivity. There is a range of impurity concentration where there is gapless superconductivity.

Russian physicists have made many further advances in the theory of superconductivity. To mention just one, making use of Migdal's theorem, G. M. Eliashberg gave a pair of coupled integral equations to derive the energy dependence of the pair potential and of the self-energy from the basic electron-phonon and Coulomb interactions. These equations since have been used a great deal to derive the properties of strongly coupled superconductors such as lead and mercury.

X.   GINZBURG-LANDAU THEORY AND TYPE II SUPERCONDUCTORS

It was in 1950 that Ginzburg and Landau proposed their phenomenological theory of superconductivity in which a complex order parameter, $\psi(r)$, with amplitude and phase is introduced. The density of the superfluid condensate, n_S, is proportional to $|\psi(r)|^2$ and the momentum p_S to the gradient of the phase. The theory thus allowed for space variations of both n_S and p_S and for the multitude of current distributions possible in a superconductor. It was supposed that $\psi(r)$ satisfies a nonlinear Schrödinger-like equation. The theory was applied to calculations of the boundary energy between normal and superconducting phases in the intermediate state, to the calculation of critical currents, and to many other problems.

At the time this work was done, the cold war was at its height. Distribution of Russian journals was slow. I was fortunate to obtain an excellent translation of the Ginzburg-Landon article through the courtesy of Shoenberg.

One of the most important applications is Abrikosov's theory of type II superconductors. In 1953, he derived a periodic solution of G-L equations which we

now recognize as a periodic array of quantized vortex
lines threading the superconductor. He applied the
theory to account for Shubnikov's experimental results
on type II superconductors and derived expressions for
H_{c1} and H_{c2} in terms of the parameters of the theory.

Abrikosov was then a young research worker in
Landau's group. The calculation was rather abstract,
and Landau did not appreciate the physical significance
of the results until Feynman's ideas on quantized vortex
lines in liquid helium were made known in 1956. Landau
saw the analogy, and then allowed Abrikosov to publish
his results that same year, three years after the work
was actually done. Even so, his paper was rather lost
sight of in the flurry over the microscopic theories
which started appearing in 1957. It now forms the basis
for our understanding of type II superconductors.

Renewed interest in type II superconductors was initiated by the discovery by J. E. Kunzler and associates
at the Bell Telephone Laboratories in 1961 that Nb_3Sn
will withstand very high magnetic fields, up to 200,000
gauss. It was this discovery that made the construction
of present day high field superconducting magnets possible. At first it was thought that an explanation could
be found in Mendelssohn's sponge model, according to
which superconductivity could remain at high fields in
narrow filaments threading the structure. It was B. B.
Goodman, then at Grenoble, who pointed out that Nb_3Sn is
a type II superconductor and the relevance of Abrikosov's
theory. Since then, a great deal of research has been
done on pure niobium and other carefully prepared type
II superconductors which have verified the theory.

One of the most active groups in this study from
both a theoretical and experimental standpoint has been
that of P. G. de Gennes and coworkers at Orsay, France.
To treat problems in which the gap parameter, $\Delta(r)$,
varies in space, they used the Bogoliubov self-consistent
field approach. One of their discoveries is the prediction that superconductivity can exist in a thin surface layer well above H_{c2}, later confirmed experimentally.

There has recently been much interest in extensions
of the Ginzburg-Landau equations which take into account
time changes of the order parameter. Among other applications, time-dependent versions can be used to explain
fluctuations of Δ above T_c that give rise to an increased
conductivity, called paraconductivity, above that of the
normal metal.

XI. TUNNELING

One of the most powerful methods for investigating superconductors is by means of electron tunneling. The discovery of the effect by I. Giaever and the subsequent prediction of supercurrent flow through a tunnel junction by B. D. Josephson show that youth still has a place in physics research. Both of these very important discoveries were made while they were still graduate students.

The story of Giaever's discovery has been described by R. W. Schmitt in an article in Physics Today. Giaever was born and educated in Norway as a mechanical engineer. He migrated to Canada where he joined the Canadian General Electric Company. Later, in 1956, he came to Schenectady on an advanced training program, and was assigned to the General Electric Research Laboratory for six months. While there, he became interested in solid state physics and decided to switch fields and become a physicist. He began doing tunneling measurements between normal metals with John Fisher. To further his education in physics, he took advanced courses at a nearby college, where he learned about superconductivity and the energy gap. He thought that the energy gap might show up in the tunneling characteristic. Since the gap corresponds to a few millivolts, it should be in an accessible range. In the spring of 1960, he observed a striking effect in an aluminum-oxide-aluminum junction. At low temperatures he found little current flow until the voltage was sufficient to overcome the gap. Tunneling between a normal metal and a superconductor gives the density of states in energy in the superconductor and information about the phonon induced interaction. Giaever recognized that tunneling between superconductors with different gaps should have a negative resistance characteristic, and this was later observed experimentally. The discovery of tunneling opened up a whole new field of research.

At the time Giaever was doing his initial experiments, I was doing some consulting for General Electric and thus had a chance to share in the excitement. In regard to the theory of the effect, there was concern as to why the density of states should be reflected in the superconducting tunnel current, but not in the current flow into normal metals. Further, there were no coherence factors such as appear in other calculations of transition probabilities. I gave a formulation of tun-

neling appropriate to many-body systems. In a superconductor, there is a degeneracy between quasiparticle
excitations for states above and below the Fermi surface
(the energy E_k is independent of the sign of ε_k), and
this must be taken into account. Later, M. H. Cohen,
Falicov and Phillips described tunneling in terms of a
Hamiltonian with three terms:

$$H = H_R + \dot{H}_L + H_T , \qquad (10)$$

where H_R describes electrons on the right of the tunnel
junction, H_L on the left and H_T has matrix elements for
transitions from right to left and left to right. This
Hamiltonian formalism has since been used a great deal.

One of those who read about this work was Josephson,
who was then a graduate student of Pippard at Cambridge
University. In 1962, he used the tunneling Hamiltonian
to show that a supercurrent should flow through a tunneling barrier separating two superconductors, and that
this current flow should be proportional to the sine of
the phase difference between the superconductors on
either side of the junction. If a voltage V is applied,
the phase difference should increase as ωt, where

$$\omega = 2eV/\hbar , \qquad (11)$$

giving an alternating current of frequency ω. With V of
the order of millivolts, ω is in the microwave range. An
applied microwave field should beat against this signal,
giving steps in the d.c. characteristic where the current
varies while the voltage remains constant. All of these
predictions were subsequently observed; the first observation of supercurrent flow was by P. W. Anderson and
J. Rowell at the Bell Telephone Laboratories and of the
steps in the current-voltage characteristic by S.
Shapiro. Since all that is required are measurements
of voltage and of frequency, the Josephson effect has
been used to make a very precise measurement of e/h, and
thus has contributed importantly to our knowledge of the
basic physical constants.

An important aspect of Josephson's theory is that
it established the relevance of phase as a variable describing supercurrent flow. Phase plays the same sort
of role for supercurrent flow that voltage does for flow
in normal metals. While the concept of phase goes back
at least to the Ginzburg-Landau theory, its importance
had not been fully recognized. In a quantum mechanical
sense, phase is a conjugate variable to particle number.

Not long after his first note appeared, I met
Josephson at an international meeting on low temperature
physics held at Queen Mary College, London. I had expressed skepticism about his results, since the formalism implicitly assumed that pairing would extend through
the tunnel junction, and I felt that a solution of the
Gor'kov equations in which the states that are to be
paired are determined self-consistently would show a more
rapid drop than assumed. We had some discussions at the
meeting, and further talks when I rode up with him on the
train to Cambridge after the meeting. I remained skeptical, but Josephson was convinced that he was right, and
he certainly was.

Aside from my doubts about formalism, I did not have
a good appreciation of the physical significance of phase
that was developed by Josephson in cooperation with
Anderson and others. There is a coupling energy between
the right and left sides of the junctions proportional to
the cosine of the phase difference, and the Josephson
current is simply related. Josephson later derived the
tunneling current by Green's function methods showing
that the effect I was worried about, while present, is
small.

There is little doubt that others doing experiments
on tunnel junctions unsuspectingly had observed Josephson
tunneling prior to Anderson and Rowell, but thought that
the current flow observed with vanishing voltage were
the results of shorts. One way to distinguish is to
apply a small magnetic field parallel to the plane of
the junction, which causes the phase to change along
the junction in a direction perpendicular to the field
and a consequent decrease in current from phase cancel-
lation. Josephson tunneling is sensitive to the magnetic
field; a short is not. A plot of the maximum current
versus magnetic field is similar to that of a single slit
diffraction pattern. The current is a minimum when one
flux quantum passes through the barrier region.

Many beautiful experiments have been done which show
that superconductors do exhibit quantum effects on a
macroscopic scale and which illustrate phase interference
effects. The first of these was the demonstration of
flux quantization by Deaver and Fairbank and by Doll
and Näbauer. They verified a prediction of F. London
that the flux threading a superconducting loop should
be quantized. The observed flux quantum, $hc/2e$, is half
that suggested by London, a result of pairing. Many
subsequent experiments have been done with Josephson

tunnel junctions. A group at the Ford Scientific Laboratories, R. C. Jaklevic, J. J. Lambe, J. E. Mercereau
and A. H. Silver, have demonstrated phase interference
effects with two Josephson junctions in parallel, analogous to double slit diffraction patterns. The effect can
be used to make an extremely sensitive detector for voltage or current, and thus has had wide application.

An important application of tunneling is to derive
very important information about the electron-phonon
interaction and thus obtain a very detailed confirmation
of the pairing theory. From the current-voltage characteristic, for tunneling from a normal metal into a
superconductor, one can get the energy dependence of the
complex pair potential, $\Delta(E)$. Since $\Delta(E)$ is all that is
required for the Green's function of the superconductor,
one can determine all of the many equilibrium and transport properties from the tunneling data, and good agreement with experiment is found. From the Eliashberg integral equations, one can derive $\Delta(E)$ from an assumed
phonon density of states, $F(\omega_q)$. These enter in the
theory in the combination $\alpha_q^2 F(\omega_q)$. This was first done
by Schrieffer and two associates, D. J. Scalapino and
J. W. Wilkins, who calculated $\Delta(E)$ for lead to compare
with experiment. Later W. L. McMillan worked out a
computer program to derive $\alpha_q^2 F(\omega_q)$ directly from the
tunneling data. Also needed is a parameter that is a
measure of the effective Coulomb interaction, which
varies slowly and does not seem to be particularly
structure sensitive.

For simple metals, such as aluminum and lead, it is
possible, with considerable effort, to calculate α_q
and $F(\omega_q)$ directly from first principles, and thus to
calculate T_c. More simply, one can determine $F(\omega_q)$
from neutron scattering data and estimate $\alpha_q^2 F(\omega_q)$ to
compare with that derived from tunneling data. This
has recently been done for a transition metal, tantalum,
confirming the electron-phonon mechanism for this case.

Since the coefficient, α, in the isotope effect,
$T_c \sim M^{-\alpha}$, departs considerably from 1/2 for many transition metals, some had thought that something besides
electron-phonon interactions would be required to account
for the superconductivity in the transition metals, but
this does not seem to be the case. Departures of α
from 1/2 for the transition metals are to be expected,
because the Coulomb interactions scale differently than
phonon interactions.

XII.   SUPERCONDUCTING MATERIALS

Superconductivity is a fairly common phenomenon.
A large fraction of the metallic elements and many
intermetallic compounds and alloys become superconducting
at low temperatures. Many of the elements become superconducting only at high pressures. The leader in the
discovery of new superconducting materials has been
B. T. Matthias, who with his associates has found a
very large fraction of the known superconductors. Of
German origin, Matthias was educated in Switzerland and
came to the USA in 1949 to work at the Bell Telephone
Laboratories. His interest at that time was in ferro-
electricity and he had discovered many new ferroelectric
materials. He went to the University of Chicago in
1950. It was E. Fermi who suggested to him that he
apply his talents to superconducting materials. A low
temperature program had been initiated at Chicago by
J. K. Hulm, newly arrived from Cambridge University.
Matthias and Hulm joined forces and began an exploration
of superconducting materials.

Matthias returned to Bell Telephone Laboratories
in 1953, where he initiated a program of low temperature
research which has since been extremely productive. One
of his principal associates there was T. Geballe, who
had obtained his Ph.D. under Giauque at Berkeley. Hulm
went to the Westinghouse Laboratories, where as Associate
Director, he has among other things been active in
directing research in superconductivity and superconducting materials. Matthias is now at the University
of California at San Diego and Geballe is at Stanford
University, both quite active in research on superconducting materials.

Part of the motivation for the search for new
materials is to try to understand the systematics and
thus get a deeper understanding of the factors that
determine the critical temperature and other parameters
of superconductors. Another is to discover materials
with unusual properties, particularly high critical
temperature and other parameters of superconductors.
Matthias has developed some empirical rules involving
electron/atom ratios and atomic volumes that have been
very useful as a guide for estimating critical temperatures and finding new materials.

The microscopic theory has played a relatively minor
role in this search for new materials. In part, this
is because there is a rough law of corresponding states;

most properties can be described in terms of a small
number of parameters, the most important of which are
the density of states in energy, the Fermi velocity and
the effective interaction that determines T_c. The
first two of these involve the normal state, only T_c
involves the superconducting state. The effective interaction from phonon and Coulomb effects which determines
T_c is very difficult to estimate from first principles,
although as we have mentioned this has been done in a
few simple cases.

Most physicists feel that the maximum possible
critical temperature from a phonon induced interaction
is probably not a great deal higher than the $21^{\circ}K$
observed in a compound of niobium, germanium and aluminum. A possible exception is a metallic form of
hydrogen, which according to theory should be formed
at very high pressures ($>10^6 Kg/cm^2$). Calculations
indicate that it may be superconducting at room temperature. One of the most challenging problems in
superconductivity is whether there is another mechanism
for getting an attractive interaction between electrons
which might operate at temperatures considerably higher
than the present maximum of $21^{\circ}K$.

A particularly interesting class of materials are
semiconductors that can become superconducting when
doped into the metallic range of conductivity. Since
the carrier concentration can be varied, one has an
additional variable to compare theory and experiment.
Polar semiconductors with a large electron-phonon
interaction are favorable. Some examples are GeTe,
SnTe and $SrTiO_3$. The person who has contributed most
to the theory is M. L. Cohen, now at Berkeley.

XIII. OTHER DEVELOPMENTS

Research in superconductivity expanded rapidly
after 1957 in both theory and experiment. We can do no
more here than indicate some of the more important
areas in which research has been done. A treatise published recently, edited by R. D. Parks, consists of two
large volumes of review articles covering various areas
of research. Many other books have appeared that deal
with more limited aspects of superconductivity.

The rapid expansion of efforts after 1957 is indicated by the attendance at international conference
on superconductivity. At Cambridge University in 1959

there were about 90 participants, at the IBM conference in 1961 about 150, and at Colgate University in 1963, about 350, roughly a two year doubling time. The rate of increase leveled off after that; there were about 400 at the Stanford conference in 1969.

Many experiments have been done to check in quantitative detail predictions of the theory in regard to various equilibrium and transport properties. Precise measurements of the critical field were used to determine the free energy difference between normal and superconducting states and thus the thermal properties. Departures from the law of corresponding states for strong-coupling superconductors such as lead and mercury have been accounted for by a combination of theory and experiments on tunneling characteristics.

Small departures from the empirical law (7) for the temperature dependence of the penetration depths predicted by the theory at low reduced temperatures were confirmed by experiment. Further experiments on the surface impedance as a function of frequency showed that the theory of the electromagnetic properties was able to account for the results over a wide range of temperature, frequency and coupling constants. The most complete check of the theory is that of J. R. Waldrum in experiments done at Cambridge University.

Areas where initially there were thought to be discrepancies between theory and experiment have gradually been resolved. For one example, the microscopic theory that the Pauli spin susceptibility should vanish as $T \rightarrow 0$, but the Knight shift does not. It is now recognized that only part of the Knight shift is associated with the Pauli susceptibility, the rest is due to spin-orbit effects.

It was at first thought that a precursor in the absorption observed by O. Ginsberg and Tinkham at frequencies just below the gap might be associated with collective excitations. However, calculations indicated that any such precursor absorption should be down by at least a factor of ten below that observed. Further experiments on well defined specimens show no precursor, in agreement with theory. There is thus, unfortunately as yet no experimental evidence for collective excitations.

Both theory and experiment have confirmed various aspects of thermal conductivity and of ultrasonic

attenuation. In the original theory, it was assumed
that in calculating ultrasonic attenuation for longitudinal waves, the mean free path was much greater than
the wave length of the sound. It was further assumed
that screening of the ion motion is the same in superconducting and normal states. Subsequent work of T.
Tsuneto showed that the simple law $\alpha_s/\alpha_n = 2f(\Delta)$ applies
more generally, for an arbitrary m.f.p., and also for
the temperature dependent part of the attenuation of
transverse waves observed below an abrupt drop near T_c.
It even applies to strong coupling superconductors, as
shown by V. Ambegaokar. The thermal conductivity is
dependent on the relative role of impurity and phonon
scattering of the electrons, and the theory is in good
accord with experiment when this is properly taken into
account.

The things that we have discussed so far are based
mainly on the original form of the theory in which it was
assumed that the superconductor is homogeneous and the
pair potential independent of position. There are many
important problems in which Δ varies in space or time
or both. Such problems can be treated by the powerful
Green's function methods. A space dependent $\Delta(r)$ can
also be treated by a generalized pairing equations
introduced by Bogoliubov, in which the wave functions
of the paired states are determined self-consistently.
With such equations, one can discuss the structure and
motion of vortex lines in type II superconductors,
proximity effects, the boundary between normal and superconducting regions in the intermediate state, and
related topics. They were used by De Gennes to predict
in advance of observation the existence of surface
superconductivity in type II superconductors in magnetic
fields higher than the upper critical field.

While current research in superconductivity covers
a broad range of topics, one of the most active is that
involving effects of fluctuations of the pair potential,
Δ. Because of thermal fluctuations, $<\Delta^2>$ differs from
zero above T_c, giving significant changes in the
electrical conductivity and in the magnetic susceptibility. Fluctuations can also give a finite rather than
infinite d.c. conductivity below T_c. Because the
fluctuating volumes are smaller, larger effects are
observed in thin films than in bulk specimens. Many
groups, both experimental and theoretical, have been
involved with the very intriguing problems in this
area.

XIV. APPLICATIONS

A major stimulus to research in superconductivity comes from applications that are beginning to appear. Annual meetings on applied aspects are creating a great deal of interest. So far, the applications have been to other areas of science rather than to the general consumer market. Serious study is, however, being given to the use of superconductors in the generation and transmission of electrical power, to computer components, and to other areas. Superconducting magnets are now commonplace in laboratories throughout the world. If fusion energy ever becomes a reality, it will be done with superconducting magnets to contain the plasma.

In high energy physics, superconducting resonant cavities are being used in linear accelerators to permit continuous rather than pulsed operation. The pioneer work and largest installation of this kind is at Stanford University, where a 500 foot superconducting linear accelerator is under construction, in which pure niobium is used in the cavity walls. Other applications are to beam bending and focusing magnets and to bubble and spark chamber magnets. Very large bubble chamber magnets have been built at Argonne and at Brookhaven.

Also at Stanford, under the direction of W. M. Fairbank, several other large and difficult experiments making use of superconductors are underway. One makes use of a superconducting gyroscope in a space satellite for a test of general relativity. In another, the principle of flux quantization is being used to make a volume completely free of a magnetic field.

It is still too early to tell whether or not superconductivity will be important in the electric power industry. Considerable research is being done, particularly in England, where a large homopolar generator has been built with a superconducting magnet. Superconducting transmission lines may be feasible if it is necessary to go underground in any case, or if very large amounts of power are to be transmitted. Since superconductors have a.c. losses, the problems are less severe for d.c. rather than a.c. transmission.

Much research has been done on superconducting computer components, which can be used for logic as well as memory. So far, it has been difficult to compete with the rapid advances in semiconductor technology.

Superconductors may have some advantages for very large
memory arrays.

In a completely different area, the Josephson effect
has made possible the design of extremely sensitive
detectors for current, voltage or magnetic field. SQUIDS
(superconducting quantum interference devices) are now
being used in many laboratories. The Josephson effect
can also be used as a voltage standard, since it can be
used to convert frequency to voltage. Active programs
are underway in standards laboratories in several
different countries.

XV.   INFLUENCE ON OTHER AREAS OF SCIENCE

Not long after the microscopic theory was discovered, Pines, together with A. Bohr and B. R. Mottleson,
suggested that the pairing energy in nuclei has a similar origin to the pairing energy in superconductors,
and that similar mathematical methods could be used in
the nuclear problem. This concept has been developed
considerably during the past decade, with superconductive pairing playing an important role in problems
of nuclear structure.

More recently, it has been suggested by D. Pines,
G. Baym and C. Pethick and by others that the neutrons
and protons in the interior of neutron stars may be
superfluid. Evidence comes from changes in the period
of pulsars following a sudden speed up.

Concepts from superconductivity have aided in
understanding the closely related phenomenon of super-
fluidity in liquid helium. Effects analogous to the
Josephson effect are observed in flow through a small
orifice connecting two chambers.

Following initial work of Nambu, attempts have been
made to extend some of the basic concepts to understand
better the elementary particles of high energy physics.
There has been considerable discussion of the possible
role of degenerate vacuum states. In a superconductor,
the degeneracy is in the phase.

One of the most important applications is the use
of the Josephson effect to measure accurately e/h. No
one would have guessed a few years ago that superconductivity would give the most precise measurement of
these fundamental constants. By measuring Josephson

interference effects in a rotating system, it may be possible to measure e/m_e, and thus the electron mass. Further applications of effects of macroscopic quantization may be possible in the future.

XVI. CONCLUDING REMARKS

In this review, I have attempted to give a picture of how research in superconductivity has grown over the years and to give the origins of some of the key ideas. It is written from the personal point of view of an active participant, and can by no means be considered an unbiased account. Further, in order to keep the article within reasonable bounds, a selection is often arbitrary. I apologize to all those whose contributions have been omitted or slighted. The developments of the past few years have been sketched only in broad outline.

In the past dozen exciting years, the field of superconductivity has matured, with the fundamentals well established. Rather than a separate phenomenon, it now is incorporated into the general body of solid state physics. Superconductivity is studied for the detailed information it gives about electron-phonon interactions, fluctuations in the neighborhood of phase transitions, magnetic scattering, and other effects important for solid state physics in general.

There is increasing concern with trying to understand the properties of particular materials. While there do not appear to be any discoveries comparable in importance to the Josephson effect in the offing, if there is one it would be the establishment of a mechanism for superconductivity other than the electron-phonon mechanism.

It is likely that much further research in superconductivity will be motivated by the various applications that are appearing. This involves both material properties and geometrical arrangements. There is a rich field in the many remarkable and varied aspects of macroscopic quantization as well as in applications that make use of the absence of resistance.

BOOKS ON SUPERCONDUCTIVITY AND RELATED SUBJECTS

A. A. Abrikosov, L. P. Gorkov, and I. E. Dzaloshinski,
Methods of Quantum Field Theory in Statistical Physics
(Prentice-Hall, Inc., Englewood Cliffs, New Jersey,
1963), translated by Richard A. Silverman.

John M. Blatt, Theory of Superconductivity (Academic
Press, London, 1964).

N. N. Bogoliubov, V. V. Tolmachev, and D. V. Shirkov,
A New Method in the Theory of Superconductivity (Consultants Bureau, Inc., New York, 1959).

Morrel H. Cohen, editor, Superconductivity in Science
and Technology (University of Chicago Press, Chicago,
1968).

P. G. De Gennes, Superconductivity of Metals and Alloys
(W. A. Benjamin, Inc., New York, 1966), translated to
English by P. A. Pincus.

David Fishlock, editor, A Guide to Superconductivity
(Macdonald and Co. Ltd., London, 1969).

C. G. Kuper, Theory of Superconductivity (Clarendon
Press, Oxford, 1968).

Cecil T. Lane, Superfluid Physics (McGraw-Hill Book
Company, Inc., New York, 1962).

Fritz London, Superfluids (John Wiley and Sons, Inc.,
New York, 1950), Volume 1.

E. A. Lynton, Superconductivity (Methuen, London, 1969).

Vernon L. Newhouse, Applied Superconductivity (John
Wiley and Sons, Inc., New York, 1964).

R. D. Parks, editor, Superconductivity (Marcel Dekker,
Inc., New York, 1969), Volumes 1 and 2.

G. Rickayzen, Theory of Superconductivity (Interscience
Publishers, London, 1965).

D. Saint-James, G. Sarma and E. J. Thomas, Type II
Superconductivity, (Pergamon Press, Oxford, 1969).

J. R. Schrieffer, Theory of Superconductivity (W. A.

Benjamin, Inc., New York, 1964).

D. Shoenberg, Superconductivity (Cambridge University Press, 1952).

P. R. Wallace, Superconductivity (Proceedings of the Advanced Summer Study Institute, McGill University, Montreal) (Gordon and Breach, Science Publishers, Inc., New York, 1969), Volumes 1 and 2.

OTHER SELECTED REFERENCES

P. W. Anderson, The Josephson Effect. (Chap. I, p. 1-43 in Progress in Low Temperature Physics, Vol. V. C. J. Gorter, ed. North Holland Pub. Co., Amsterdam, 1967.)

J. Bardeen and J. R. Schrieffer, Recent Developments in Superconductivity, (Physics Today, 16, No. 1. p. 19, 1963).

J. Bardeen, Advances in Superconductivity, (Physics Today, Vol. 22, No. 10, p. 40 (1969)).

E. F. Hammel, Soviet Low Temperature Physics, (Physics Today, Vol. 20, No. 8, p. 21, 1967).

B. D. Josephson, Supercurrents Through Barriers, (Advances in Physics Vol. XIV, 419 (1965)).

Proceedings, International Conference on the Science of Superconductivity, Colgate University, (Rev. Mod. Phys. 36, 1 (1964).

R. W. Schmitt, The Discovery of Electron Tunneling into Superconductors, (Physics Today, Vol. 14, No. 12, p. 38, 1961).

R. N. Taylor, W. H. Parker, and D. N. Langenberg, Determination of e/h Using Macroscopic Quantum Phase Coherence in Superconductors, (Rev. Mod. Phys. 41, 375 (1969)).

WILLIS E. LAMB, Jr.

PHYSICAL CONCEPTS IN THE DEVELOPMENT OF THE MASER AND LASER

Willis E. Lamb, Jr.

Yale University

New Haven, Connecticut

I.   INTRODUCTION

Over the last twenty years, a new branch of applied physics called Quantum Electronics has developed. Lasers, generators of highly coherent electromagnetic radiation, play a central role in quantum electronics. The word "laser" was derived by a letter change from "maser" which is itself an acronym for "Microwave Amplification by Stimulated Emission of Radiation". The maser was invented in 1951 by Charles H. Townes, then of Columbia University, and reduced to practice in 1954 by James P. Gordon, Herbert J. Zeiger and Townes. The maser principle was extended in 1958 to optical frequencies by Townes and Arthur H. Schawlow, and the first operating laser was built by Theodore H. Maiman in 1960. At first, the phrase "optical maser" was used, but this has largely been replaced by the simpler term "laser" to describe a device which generates light instead of microwaves.

There are, by now, many different forms of lasers: gas, ion, jet, shock wave, liquid, solid, dye, flame, chemical, semi-conductor junction, etc. lasers. They, and their microwave predecessors, have produced electromagnetic radiation ranging from radio frequency to the vacuum ultraviolet.

This radiation is highly monochromatic. In some cases, enormous powers can be produced. Because of its high monochromaticity, such light can be focused into a

very small region to produce an enormous energy density,
or it can be sent in a beam detectable on Earth after
striking a reflector placed on the Moon.

The listing and description of all applications of
lasers would fill a whole book. New developments in
this field have been coming at an increasing rate in the
last few years. A recent issue of Physics Today (March,
1971) contains a good summary of some presently realized
applications, as well as forecasts of future possibilities.
The extent of development of a large industry based on
the laser can be judged by examining current issues of
the trade journal Laser Focus.

The purpose of this article is not to deal with the
very important matters referred to in the preceding
paragraphs, but instead to review critically certain
aspects of the history of physics which bear on the development of the laser. To understand the mechanism
of the laser one needs such branches of physics as
mechanics, acoustics, electricity and magnetism, thermodynamics, statistical mechanics, relativity theory,
spectroscopy, atomic physics, and the quantum mechanical
theory of both matter and the radiation field. The
threads of the story which led to the invention of the
laser pass through or near almost all parts of physics.
The pre-history of the laser illustrates the way in which
scientific research of little or no apparent practical
utility has, in a wholly unforeseeable and unexpected
way, led to development of most useful and important
devices.

It is difficult to find a convenient starting point
for the discussion of our subject. We could go back to
the beginning with the biblical "And God said, 'Let
there be light' and there was light." The first light
was clearly not that of the sun, moon or stars, for these
objects were made later. Genesis does not indicate
whether the radiation was coherent or incoherent, and
it might well have been laser light. The light was presumably corpuscular rather than wave-like in character
since "God divided the light from the darkness" which would
seem to rule out diffraction phenomena.

For many centuries, physicists have asked and tried
to answer the questions: "What is matter?" "What is
light?" and "How do light and matter interact with one
another?" The questions are still valid today, because
with all our knowledge, we have only pushed the frontiers

of ignorance back. There is a wide area of explored
territory, and we have enough knowledge to be able to
design many devices for our use and amusement. However, it is very risky to try to foresee the future, and
to say that we now know all that we "need to know".
For the development of the laser, more detailed questions
had to be considered, such as "What does quantum mechanics mean?" "What are photons?" and "What is stimulated
emission of radiation, and what are its properties?"

In the long history of radiant energy, two views on
the nature of light have competed - the notions: (1)
that light consists of particles and (2) that light consists of waves. With the development of quantum
mechanics during the last half-century, it is now possible to say that both statements have some validity,
but that in general neither is right. There is, however, even among very competent physicists, still a
surprising lack of understanding of the nature of light
as clearly given by the present theory, called the
Quantum Theory of Radiation, or Quantum Electrodynamics.
The legacy of confusion inherited from previous generations
of physicists is not easily overcome. The richness of
this legacy can be seen in the history of the subject.

In order to bring the story efficiently up to modern
times, I will give a chronological listing of major
developments in optical science which bear on the development and understanding of the laser. Each item will
consist of a date, the name (or names) of the contri-
butor(s) and a short statement of his accomplishment.
There must inevitably be some errors of omission or
inclusion. With a few exceptions, fuller accounts of the
historical developments before 1927 can be found in the
two volumes of E. T. Whittaker's History of the Theories
of Aether and Electricity. In appropriate cases I have
added some remarks on the significance of the work for
the development of optical concepts. Some of the
listed items will lie outside the field of optics. Most
of these will involve advances in electricity and
magnetism with special importance for us, but a few will
represent landmarks of other branches of physics
required for the development of the whole subject.

Section III gives a list of some books which bear
on the subject. The article continues with two essays
entitled: Section IV. What are photons? Section V.
What is stimulated emission? and with Section VI. Con-
cluding Remarks.

II. CHRONOLOGICAL LISTING

Fifth century B.C. Empedocles of Agrigentum. First corpuscular theory of light, finite speed.

Democritus (460-370 B.C.). (following Leucippus). Beginnings of Atomic theory.

Plato (427-347 B.C.). The Platonists thought of vision as a divine fire with a stream of particles emitted by the eye, combining with solar rays at the object seen, and returned to the eye to give it perception.

Aristotle (384-322 B.C.) Defined light as an activity in a transparent medium. The medium acquired an instantaneous quality from the luminous object.

300 B.C. Euclid. Law of reflection, and approximate law of refraction.

Lucretius (97-54 B.C.) (following Democritos). Atomic theory.

Lucius Seneca (4 B.C.-65 A.D.). Knew that colors were produced from white light passing through a prism.

70 A.D. Claudius Ptolemy of Alexandria. Ratio of incident and refracted angles is constant for an interface of given materials.

Ibn al-Haythan (Alhazen), died c. 1039 A.D. One of the few dissenters, before the twelfth century, from the Aristotelian view of instantaneous activity. Finite speed of light, different speeds in different media. Author of important treatise on optics. Refuted feeler-emission theory of the Platonists.

1583 A.D. Galileo Galilei (1564-1642). Observed isochronous pendulum. More like the damped electromagnetic oscillations of Hertz than like the continuous wave oscillators of Rayleigh, De Forest and Townes. First attempt to measure speed of light, signals sent to and fro.

1611. Johannes Kepler (1571-1630). Treatise Dioptrics discussed refraction, opposed action at a distance.

1611. Franciscus Maurolicus (1495-1575). Publication of the Photismi de lumine. Translated by H. Crew. (MacMillan, New York, 1940).

1637. René Descartes (1596-1650). Three volumes dealing with elaborate optical theory: Discours, Dioptriques, Météores. Mechanical model for aether. "derivation" of Snell's law. Beside Whittaker, see A.I. Sabra, Theories of Light from Descartes to Newton.

1641. Pierre Gassendi (1592-1655). Revived atomism.

1644. Evangelista Torricelli (1608-1655). Showed that light propagated through a vacuum.

1661. Pierre Fermat (1601-1665). Fermat's principle. Controversy with Descartes over derivation of Snell's law.

1665. Francesco Grimaldi (1613-1703). Discovered diffraction.

1665. Robert Hooke (1635-1703). Suggested light might be a transverse wave phenomenon. Wrote Micrographia, was aware of colors of thin films, illumination in shadow.

1666. Isaac Newton (1643-1727). Colors arising from dispersion in a prism. White light as a mixture of all colors. (First published in 1672).

1669. Erasmus Bartholin (1625-1698). Double refraction of calcite crystals. Snell's law for ordinary ray.

1672. Ignace-Gaston Pardies (1636-1673). Waves. Foresaw finite velocity of light and aberration.

1675. Olaf Roemer (1644-1710). First measurement of the velocity of light, from eclipses of Jupiter's moons.

1678. Christian Huygens (1629-1695). Traité de la lumiére...(publication delayed until 1690). Favored wave theory because of non-interference of intersecting light beams, and perhaps from observations of waves on Dutch canals. His waves consisted of pulses, and were not waves of definite wave length. Noted synchronization of loosely coupled pendulum clocks.

1687. Newton. First edition of Principia.

1699. Nicolas de Malebranche. (1638-1715). Color of monochromatic light depends on period of vibration, and brightness of light increases with wave amplitude.

1704. Newton. Opticks (first edition). Space permeated by an elastic medium (aether) capable of propagating vibrations similar to sound waves in air, but with far greater velocity. These waves were not identified with light for two reasons: (1) rectilinear propagation, and (2) polarization of light. (Later developments of wave theory by Young and Fresnel removed these difficulties.) The aether, not necessarily homogeneous, filled the pores of all material bodies. Light is of a different kind from aether waves, is propagated from luminous bodies, perhaps but not necessarily corpuscular. Light interacts with the aether and its waves. Refraction, reflection and inflection (i.e., diffraction) arise from coupling of light with layers of aether at interfaces. The colors of thin films are caused by "fits of easy reflection and transmission: at surfaces. These are produced by interaction with the aether waves accompanying the light particles from previously encountered interfaces. The possibility of having simultaneous refraction and reflection at a single surface was attributed to fits produced by the aether waves sent out from the light source. The velocity of light

in optically dense media was supposed to be greater
than in vacuum, while the reverse was true on the wave
theory of Huygens. Some recent writers have seen in the
fits of easy reflection and transmission a prevision of
the quantum theory, but this view has little justification.

    Roger Cotes (1682-1716). First Plumian Professor at
Cambridge. Close associate of Newton. Strongly corpuscular in viewpoint. He stated clearly that the aim
of theoretical physics is to predict the future, and
that a consideration of interphenomena (i.e., things
going on between the start and finish of the predictive
enterprise) should have no place in the theory.

    1720. Willem 'sGravesande (1688-1742). Follower
of Newton. Wrote influential book on heat and light
Elements of Mathematical Physics Confirmed by Experiment.
Corpuscular theory.

    1728. James Bradley (1692-1762). Discovery of
aberration.

    1730. Chester M. Hall (1703-1771). Achromatic lens.

    1738. Robert Smith (1689-1768). Relative of
Cotes. Succeeded him in Plumian chair at Cambridge.
Wrote A Complete System of Optics, in 4 volumes. The
followers of Newton were strongly oriented in favor of
the corpuscular theory and played down Newton's vague
notion of fits of easy reflection and transmission.
Unfortunately, they had nothing better to offer in
explanation of some basic optical phenomena.

    1744. Pierre de Maupertuis. (1698-1759). Principle
of least action, favored corpuscular theory.

    1745. Daniel Bernoulli (1700-1782). Solution of
vibrating string problem in terms of a trignometric
series.

    1746. Leonhard Euler (1707-1783). Favored wave
theory, stressed similarity of light and sound. Argued
that emission of particles would cause loss of mass,
unlike emission of waves. Same aether for electricity,
light and gravitation.

    1752. Benjamin Franklin (1706-1790). Favored
wave theory. Law of conservation of electric charge.

    1763. Ruggiero G. Boscovich (1711-1787). Built
up extensive theory of natural phenomena based on action
at a distance.

    1785. Charles Coulomb (1763-1806). Inverse square
law of electrostatics.

    1792, 1809. Pierre Prévost (1751-1839). Formulated
law of heat exchange.

    1796. Henry Brougham (1778-1868). Extreme form of
emission theory. Later made outrageous attack on Thomas
Young and wave theory.

    1800. William Herschel (1738-1822). Discovered

infrared radiation.

1801. Johann W. Ritter (1776-1810). Discovered
ultraviolet radiation.

1801. Thomas Young (1773-1829). Founder of modern
wave theory of light, with explanation of interference
and colors of thin plates.

1802. William Wollaston (1766-1828). Found seven
dark lines in the spectrum of sunlight.

1803. Young. Bakerian Lecture of the Royal
Society. Fringes of shadows. Strongly attacked by
Brougham.

1807, 1822. Jean Fourier (1768-1830). Convergence
of trignometric series named after him.

1810. Etienne-Louis Malus (1775-1812). Polarization of light.

1810. Wolfgang von Goethe (1749-1832). A master of
literary style described the production of color by all
means available to a household in eighteenth century
Weimar, and explained them in a personal version of
Aristotle's view of color. No mention of wavelength.

1814-1815. Joseph Frauenhofer (1787-1826). Many
lines in solar spectrum.

1815. David Brewster (1781-1868). Polarization
by reflection. Strongly corpuscular viewpoint.

1815. Augustin Fresnel (1788-1827). Diffraction
phenomena using wave theory and Fresnel zones.

1817. Young. Transverse light waves.

1818. Fresnel. Fresnel dragging coefficient in
moving media.

1820. Hans Oersted (1777-1851). Magnetic field
produced by a current.

1820. Jean-Baptiste Biot (1774-1862) and Felix
Savart (1791-1841). Biot-Savart law for magnetic field
produced by a current.

1820, 1825. André-Marie Ampère (1775-1836).
Researches on electromagnetism.

1821. Georg Ohm (1787-1854). Ohm's law of
resistance.

1821. Fresnel. Theory of double refraction. A
model for dispersion of refractive index.

1823. Fresnel. Theory of reflection and refraction.

1826. W. Fox Talbot (1800-1877). Chemical analysis
by flame spectroscopy.

1828, 1830, 1837. William R. Hamilton (1805-1865).
Introduction of characteristic function, first in optics,
later in dynamics.

1829-1839. Louis J. Daguerre (1789-1851). Inventor
of Daguerreotype method of permanent photography, after
1822 work by Joseph H. Niepce (1765-1833).

1830, 1836, 1839. Augustin-Louis Cauchy (1789-1857). Model for dispersion. Aether as an elastic solid.

1830. Leopoldo Nobili (1784-1835). Used thermo-couple and thermopile for detection of infrared radiation.

1831. Brewster. Last gasp of corpuscular theory (until 1905). His Treatise on Optics revived fits of reflection and transmission in a vain attempt to match achievements of the wave theory. The corpuscular view prevailed at Cambridge until about 1840.

1831. Michael Faraday (1791-1867). Electromagnetic induction.

1831-1855. Faraday. Experimental Researches in Electricity.

1833. William H. Miller (1801-1880). Absorption lines.

1834, 1835. Hamilton. Hamilton's principle in dynamics.

1835, 1837, 1839. James MacCullagh (1809-1847). Aether as an elastic solid.

1837, 1842. Franz Neumann (1798-1895). Aether as an elastic solid. Deduced electromagnetic induction from Ampere's law and dynamical principles. Vector potential.

1837, 1839. George Green (1793-1841). Aether as an elastic solid. Enriched mathematical physics.

1842. Christian Doppler (1803-1853). Doppler effect.

1845. George Stokes (1819-1903). Explained away some obvious difficulties with solid aether theories.

1846. Wilhelm Weber (1804-1890). Proposed law of force between moving charges.

1849, 1850. Léon Foucault (1819-1868) and Armand H. Fizeau (1819-1896). Velocity of light by rotating cogwheel and by rotating mirror methods. They also measured the velocity of light in air and in water, with a result strongly in favor of the wave theory.

1849, 1852. Stokes. Diffraction theory, scattering.

1850. P. Guitard. Observed that dust particles stick together when electrified.

1851. Frederick Scott-Archer. Use of wet collodion in photography, followed in 1871 by the introduction of the gelatine emulsion by R. L. Maddox.

1852. Stokes. Nature of fluorescence. Futile attempts to detect the earth's motion through the aether.

1855. Heinrich Geissler (1815-1879). Invented mercurial air pump, making possible the work of:

1858. Julius Plücker(1801-1868), on electric dis-charges in gases at low pressure.

1858. Balfour Stewart (1828-1887). Formulated principle that the radiating power of a body equals its absorbing power.

1859. Gustav Kirchoff (1824-1887). More general
theory of thermal radiation.

1859. Kirchoff (1824-1887) and Robert W. Bunsen
(1811-1899). First practical spectroscope. Also, the
reciprocity law for photography.

1859. Fizeau. Measured velocity of light in flowing water.

1859, 1873. Dmitri Mendeleev (1834-1907). Periodic
table of the elements.

1861, 1862. James Clerk Maxwell (1831-1879).
Mechanical conception of the electromagnetic field.

1862. F. P. Leroux. Found anomalous dispersion in
iodine vapor. Also discovered much earlier by Talbot,
but unpublished.

1864. Maxwell. Published <u>A Dynamical Theory of
the Electromagnetic Field</u>.

1868. Joseph Bousinesq (1842-1929). Simplified
aether model: one all pervasive aether, with embedded
matter.

1868. Andres Ångström (1814-1874). Accurate measurements of wave lengths.

1869. Maxwell. Theory of anomalous dispersion.
Cambridge Tripos examination question.

1870. Hermann von Helmholtz (1821-1894). Theory
of reflection and refraction using Maxwell's electromagnetic theory.

1871. Maxwell. Inagural lecture for Cavendish
Professorship: There was a popular opinion, in a note-
worthy quotation, that in "a few years all the great
physical constants will have been approximately estimated,
and that the only occupation which will be left to men
of science will be to carry on these measurements to
another place of decimals." But "we have no right to
think thus of the unsearchable riches of creation, or
of the untried fertility of those fresh minds into
which these riches will continue to be poured." "The
history of science shows that even during that phase of
her progress in which she devotes herself to improving
the accuracy of the numerical measurement of quantities
with which she has long been familiar, she is preparing
the materials for the subjugation of new regions, which
would have remained unknown if she had been contented
with the rough methods of her early pioneers."

An example of remarkable foresight occurs at the
end of the lecture, where Maxwell reflects on deep
questions of physics - whether matter was continuous
or discontinuous, and whether it was capable of mathematically deterministic description. (Remember that
Maxwell had developed statistical methods for the kinetic
theory of gases, but only as a mathematical convenience,

and not as a physical necessity). "The theory of atoms
and void leads us to attach more importance to the
doctrines of integral numbers...the statistical method...
which in the present state of our knowledge is the only
available method of studying real bodies, involves an
abandonment of strict dynamical principles and an adoption
of the mathematical methods belonging to the theory of
probability. It is probable that important results will
be obtained by the application of this method, which is
as yet little known, and is not familiar to our minds.
If the actual history of science had been different, and
if the scientific doctrines most familiar to us had
been those which must be expressed in this way, it is
possible that we might have considered the existence
of a certain kind of contingency a self-evident truth
and treated the doctrine of philosophical necessity as
a mere sophism." Thus Maxwell in 1871 was prepared for
ideas necessary in quantum mechanics which de Broglie,
Schrödinger and Einstein, among others, found very
difficult to accept.

1871. George J. Stoney (1826-1911) suggested using
wave numbers instead of wavelength in looking for
spectral regularities.

1872. W. Sellmeier. Theory of anomalous dispersion.

1872. Eleuthère E. Mascart (1837-1908). Futile
attempt to detect earth's motion through aether.

1872. Ernst K. Abbe (1840-1905). Exact theory of
the microscope. High quality optical instrumentation.

1873. Frederick Guthrie (1833-1886). Electroscope
discharged when brought near a hot body.

1874. Maxwell. Treatise on Electricity and Magnetism.

1876. Oliver Heaviside (1850-1925). Telegrapher's
equation.

1876. Henry A. Rowland (1848-1901). Convection
currents produce magnetic fields.

1879. Josef Stefan (1835-1893). Stefan's law of
black body radiation.

1879. David E. Hughes (1831-1900). Change of
resistance of carbon granules in microphone transmitter.

1880. Samuel P. Langley (1834-1906). Invention of
bolometer for detection of infrared radiation. Subsequent advances include pneumatic, photographic, photoconductive, photoelectric and superconducting detectors.

1881. Albert A. Michelson (1852-1931). First
Michelson-Morley experiment.

1882. Rowland. High quality diffraction gratings.

1883. Lord Rayleigh (John W. Strutt, third Baron)
(1842-1919). Theory of periodic vibrations sustained
by a steady energy source. (Earlier important qualitative
discussions in 1878 and 1879). First example of a

problem similar to those of maser and laser.

1883. Horace Lamb (1849-1934). Skin effect.

1883. George F. FitzGerald (1851-1901). Suggested that electromagnetic radiation could be produced by time varying currents. (Magnetic oscillator in contrast to arrangement later used by Hertz.)

1883-1885. Hartley, Livering, Dewar and Cornu. Various regularities in spectral wave lengths.

1884. Ludwig Boltzmann (1849-1934). Derivation of Stefan's law from thermodymanics of radiation pressure.

1884. C. Onesti. Change of resistance of metal filings in electric field.

1884. John H. Poynting (1852-1914). Energy conservation law for electromagnetic field (Poynting's vector expression power flux.)

1885. Heaviside. Similar theorem for momentum density in an electromagnetic field.

1885. Joseph Balmer (1825-1898). Balmer series for hydrogen.

1886, 1888. Heinrich Hertz (1857-1894). Generation, detection and properties of electromagnetic waves (of frequency about 75 MHz.)

1887. Hertz. Discovery of the photoelectric effect.

1888. Oliver Lodge (1851-1940). "The whole subject of radiation seems working itself out splendidly", as quoted by Condon and Shortley in Theory of Atomic Spectra.

1888. Heaviside. Magnetic field of uniformly moving point charge.

1889. Heaviside. Force on a point charge moving in a magnetic field.

1889. Hertz. Theoretical analysis of his experiments according to Maxwell's electromagnetic theory.

1890. Johannes Rydberg (1854-1919). General equation for spectral series.

1890. Edouard Branly (1844-1940). Iron filings in glass tube, used as detector of electromagnetic fields.

1892. Lodge. Light velocity not affected by motion of nearby matter.

1892. FitzGerald. Suggested FitzGerald-Lorentz contraction to explain null result of Michelson-Morley experiment.

1892, 1895. Hendrik A. Lorentz (1853-1928). Electron theory of electrodynamics of moving media. Model for dielectrics based on harmonic oscillator atoms.

1893. Joseph J. Thomson (1856-1940). Electromagnetic momentum of moving charge.

1893. Wilhelm Wien (1864-1928). Wien's displacement law for black body radiation.

1894. Lodge. Pen recording of Hertzian waves using
trembler decoherer (equivalent to tapping the coherer
in order to restore its sensitivity).

1894. Stoney. Coined word "electron", but concept
was discussed by him as early as 1874.

1895. Guglielmo Marconi (1874-1937). First sensitive
detector of Hertzian waves. Coherer (based on earlier
work by Guitard (1850), Hughes (1879) and Onesti (1884).
One mile range in 1865, 200 miles by 1901. Other detectors of electromagnetic radiation were: steel needle
resting lightly on a carbon block, a platinum wire
dipped in nitric acid (1902), cat's whiskers on galena
(1903) and silicon crystals (1906).

1895. Alexander S. Popov (1859-1905). Electromagnetic pulses generated by distant lightning flashes.

1895. Joseph Larmor (1857-1942). Electromagnetic
self-energy of a moving charge.

1895. Wilhelm K. Röntgen (1845-1923). Discovery
of x-rays.

1896. Wien. Wien's law for spectral distribution
of thermal radiation.

1896. Henri Becquerel (1852-1908). Discovery of
radioactivity.

1897-98. Charles T. R. Wilson (1869-1959). Cloud
chamber.

1897-1901. Charles Fabry (1867-1945) and Alfred
Perot (1863-1925). Fabry-Perot interferometer.

1898. Lodge. Concepts of resonance and tuning.

1898. Marie S. Curie (1867-1934) and Pierre Curie
(1859-1906). Discovery of polonium and radium.

1899. Henry Poincaré (1854-1912). Expressed
opinion that absolute motion was undetectable in principle,
only relative motion of source, medium or detector produced measurable effects.

1900. Otto R. Lummer (1860-1925) and Ernst Pringsheim
(1859-1917). Measurement of black body spectrum used
by Plank to determine h and k.

1900. Lord Rayleigh. Derivation of Rayleigh-
Jeans law for spectral distribution of thermal radiation.
(James Jeans paper in 1905).

1900. William B. Duddell (1872-1917). Arc oscillator for generation of electromagnetic waves. Pre-
ceeded by rotary spark gap, quenched sparks, and by
tuned sparks producing continuous wave oscillation.

1900. John S. Stone (1869-1943). Loose coupling
of resonant circuits.

1900. Paul U. Villard (1860-1934). Gamma rays.

1900. Ernest Rutherford (1871-1937). Exponential
law of radioactive decay.

"1900". E. T. Whittaker says in his <u>History</u> <u>of the Theories</u> <u>of</u> <u>Aether</u> <u>and</u> <u>Electricity</u>", vol. 2, ch. 3 that "At the beginning of the nineteenth century the theory of radiation was in a most unsatisfactory state."

1900-01. Max K. Planck (1858-1947). Law of spectral distribution of black body radiation. Atoms emit and absorb energy in units $h\nu$.

1901. Heinrich Rubens (1865-1922) and F. Kurlbaum. Improved measurements of black body spectrum.

1901. Reginald A. Fessenden (1866-1932). High frequency alternator.

1902. Frederick Soddy (18770-1956). Successive radioactive transformations.

1902. Lummer and Ernst J. Gehricke (1878-1960). Interference effects over distances of 2×10^6 wavelengths of green light.

1902-1903. Albert Einstein (1879-1955). Theory of Brownian motion.

1902-1903. F. T. Trouton and H. R. Nobel. Trouton-Nobel experiment to detect torque on a charged condenser, supposedly to be caused by motion through the aether. Other similar experiments by FitzGerald (before 1901), Rayleigh (1902) and Brace (1904) also gave negative results.

1902. J. Willard Gibbs (1839-1903).(Yale!) Generalized Boltzmann distribution for any dynamical system in thermodynamic equilibrium.

1902-1903. Rutherford and Soddy. General theory of radioactive transformations.

1903, 1908. Rutherford. Identification of alpha particles with helium ions.

1903. Lorentz. Maxwell's equations are invariant under a Lorentz transformation.

1903, 1910. J. J. Thomson. X-rays as "needle" radiation.

1904. J. J. Thomson. Thomson model of atom, negative electrons embedded in sphere of uniformly distributed positive charge.

1904. Charles G. Barkla (1877-1944). Polarization of x-rays.

1904. John A. Fleming (1849-1945). Vacuum diode. Use as rectifier and as detector of incoming electromagnetic radiation.

1904. Poincaré. Formulated the Principle of Relativity. The laws of physical phenomena must be the same for a "fixed" observer as for an observer who has a uniform motion of translation relative to him: so that we have not, and cannot possibly have, any means of discerning whether we are, or are not, carried along in such a motion. ...there must arise an entirely new kind of dynamics, which will be characterized above all by the rule that no velocity can exceed the velocity of light.

1905. Egon von Schweidler (1873-1948). Emission of alpha particles regarded as an intrinsically random process, of Markoff type.

1905. Einstein. Theory of relativity.

1905. Einstein. Theory of the photoelectric effect.

1905-1906. Einstein. Concept of light quanta as an independent entity in the aether, from entropy considerations. The word "photon" was coined much later by Gilbert N. Lewis in 1926.

1906. Einstein. Theory of specific heats.

1907. Hans Geiger (1882-1945) and Rutherford. Geiger counter.

1907. Arthur W. Conway (1875-1950). Before Conway, line spectra were considered to be produced by free vibrations of a complicated dynamical system. Conway got the idea of transient excited states, each of which gave radiation of a particular frequency. The whole spectrum came from a collection of atoms in different excited states.

1907. Lee De Forest (1873-1961). Invented "audion," grid added to diode to make triode vacuum tube.

1908. Walther Ritz (1878-1909). Ritz combination principle.

1909. Lorentz. Theory of Electrons.

1909. Geoffrey I. Taylor (1886-   ). Interference effects with very weak light. More accurate tests by A. Dempster and H. Batho in 1927.

1909. Ernst F. Alexanderson (1878-   ). High frequency alternator generating 10^5 Hz, running at 20000 rpm, with 300 poles on stator, air gap of 0.015".

1909. Geiger and E. Marsden. Large angle scattering of alpha particles by matter.

1909. Einstein. Fluctuations of energy density in black body radiation expressed as sum of wave-like and particle-like contributions.

1910. Harry Bateman (1882-1946) and Rutherford; also, Geiger and Bertram B. Boltwood (1870-1927). Fluctuations in radioactive used to determine Avogadro's number.

1910. William H. Bragg (1862-1942). Photoeffect by x-rays interpreted by corpuscular model.

1910. Georges Claude (1870-1960). Laid basis for neon sign industry, which was well developed by 1920.

1911. Rutherford. Nuclear atom and theory of alpha-particle scattering.

1911. Louis Dunoyer (1880-1960). First atomic beam experiments.

1911. Arnold Sommerfeld (1868-1951). Suggested that an action of $h/2\pi$ was exchanged in every elementary process. Applied to photoeffect by Sommerfeld and Debye in 1913.

1911. John W. Nicholson (1881-1955). Another precursor of Bohr's 1914 theory, quantization of angular momentum.

1911. Max Planck. In an effort to reconcile photons with electromagnetic theory, he proposed that emission occurs discontinuously in quanta, while absorption occurred continuously via classical electromagnetic theory. Radiation in transit was described by Maxwell's theory, and an atomic oscillator might have any energy value. When an atomic oscillator has an energy equal to or greater than $h\nu$ it may emit a corresponding photon, the emission being governed by the laws of chance. (Called Planck's second theory, led to zero point energy of the radiation field.) There was even a third theory. Emission and absorption by the atomic oscillators are continuous and governed by classical eletromagnetic theory. Quantum discontinuities arise only in collisions between the atomic oscillators and "free" particles. This theory was abandoned very quickly following calculations by A. D. Fokker in 1914 of the interaction of a system of electric dipole rotators in a classical electromagnetic field.

1911, 1917. Paul Ehrenfest (1880-1933). Adiabatic invariants of dynamical systems, and their necessary involvement in rules of quantization.

1912. Niels Bjerrum (1879-1958). Quantum conditions applied to diatomic molecules. Work extended by Ehrenfest in 1913.

1912. W. Friedrich and P. Knipping. X-ray diffraction by crystals, an effect suggested by Max von Laue on the basis of estimate by Sommerfeld that the wavelength was about 0.3A. This, in turn, was determined from indications that x-rays could be diffracted by wedges in observations by Haga and Wind (1899, 1902), Walter and Pohl (1908, 1909) and Koch (1912).

1912. Lee De Forest. Vacuum tube triode feedback oscillator, generating continuous wave oscillations. The U. S. Supreme Court in two decisions of 1928 and 1934 gave the patent to De Forest over three other claimants: Alexander Meissner, Edwin H. Armstrong and Irving Langmuir.

1913. Johannes Stark (1874-1957). Effect of electric field on spectral lines.

1913. J.J. Thompson. Isotopes, of neon.

1913. Max von Laue (1879-1960). Theory of diffraction of x-rays by crystals.

1913. Henry G. Moseley (1887-1915). Assignment of atomic numbers from wave-lengths of K and L x-ray lines.

1914. W. H. Bragg and William Lawrence Bragg (1890-1971). X-ray spectrometer, structure of crystals.

1914. Adriaan D. Fokker (1887-   ). Electric dipole rotators in a classical electromagnetic field.

1914. O. W. Richardson. The Electron Theory of Matter.

1914. Niels Bohr (1885-1962). Quantum theory of Rutherford atom. Stationary states determined by quantum conditions. Radiative transitions according to Ritz combination principle using Planck-Einstein relation $E_2 - E_1 = h\nu$ for emission and absorption. Abandoned all classical attempts to describe behavior of the active electron during the transition from one stationary state to another.

1914. James Franck (1882-1964) and Gustav Hertz (1887-   ). Excitation of discrete atomic levels by electronic collisions.

1915. Robert J. Strutt (Lord Rayleigh, 4th Baron.) (1875-   ). Studies on fluorescence.

1915. Sommerfeld and, independently, William Wilson (1875-1965). Quantization using action integral.

1916. Sommerfeld. Relativistic fine structure.

1916. Sommerfeld and Peter Debye (1884-1966). Theory of (normal) Zeeman effect according to Bohr theory.

1916. Karl Schwarzschild (1873-1916) and Paul S. Epstein (1883-1966). Theory of the Stark effect using Bohr theory.

1916. Einstein. Derivation of Planck radiation law assuming atoms distributed in stationary states according to the Gibbs law with radiative transitions corresponding to absorption, spontaneous emission, and induced emission. Intensity of spontaneous radiation proportional to the transition probability. Einstein also considered the atomic recoil momentum associated with emission and absorption, and showed how Maxwell-Boltzmann distribution for the atoms could be maintained by taking the momentum $h\nu/c$ of a light quantum into account. He saw that it was necessary to think of spontaneously emitted radiation as a directed process (needle radiation), and not an angularly distributed one as suggested by classical electromagnetic theory.

1916-1919. Gilbert N. Lewis (1875-1946). W. Kossel (1888-1956) and Irving Langmuir (1881-1957). Valence theory.

1918. Bohr. Correspondence Principle.

1918. Adalbert Rubinowicz (1889-   ). Selection rules related to angular momentum.

1918. G. Duffing. Monograph dealing with forced oscillations of a simple pendulum.

1919. Hendrik A. Kramers (1894-1952). Correspondence principle used for calculating intensities of spectral lines.

1920. Michelson. Measured stellar diameters by interferometry, with interference between rays over 24 feet apart. Query for believers in photons: How can such a large, low intensity, beam be made of single photons? References on the status of the coherence problem in the '20's: G. P. Thomson, Proc. Roy. Soc. 104, 115 (1923), E. C. Stoner, Proc. Camb. Phil. Soc. 22, 577 (1925) and E. O. Lawrence and J. W. Beams, Proc. Nat. Acad. Sci. 13, 207 (1927).

1920. Balthasar van der Pol (1889-1959). Theory of triode oscillator.

1920. Heinrich Barkhausen (1881-1956) and K. Kurtz. Article titled, "The Shortest Waves Producible by Means of Vacuum Tubes", describing the Barkausen-Kurtz oscillator.

1921. Otto Stern (1888-1969) and Walther Gerlach (1889-   ). Space quantization of silver atoms in an atomic beam.

1921. Albert W. Hull (1880-1966). Theory of the magnetron. Two early forms of oscillators were: (1) The negative resistance or dynatron type and (2) The transit time magnetron. The cavity resonator magnetron was invented by J. T. Randall and H. A. Boot at the University of Birmingham early in World War II.

1921, 1928. Rudolf Ladenberg (1882-1952). Discussed scattering, refraction and dispersion of light according to the old quantum theory.

1922. Louis de Broglie (1892-   ). Analysis of Planck's radiation formula into terms describing "photon molecules" of energy $sh\nu$ where s = 1, 2, 3, ..., anticipation of Bose-Einstein statistics.

1922-1923. Arthur H. Comptom (1892-1962). Compton effect, interpretation in terms of energy and momentum of light quantum.

1923. Walter Bothe (1891-1957). Gave a new derivation of Planck's distribution in terms of light quantum molecules made of s photons, s = 1, 2, 3, ... A molecule of energy $sh\nu$ interacting with an atom can excite the atom and be changed to a photon molecule of energy $(s - 1)h\nu$. A molecule of energy $sh\nu$ interacting with an excited atom can cause stimulated emission to a lower state. Bothe assumed that the emitted quantum had

exactly the same frequency and velocity as the incident
one and resulted in a photon molecule of energy (s +1)hν.
Writing detailed balance conditions for sponteneous
emission and the various induced emission processes
and absorption processes, and assuming the Gibbs distribution for the atomic states, Bothe was able to derive
the Planck distribution law.

    1923. Alfred Landé (1888-   ). Landé splitting
factor in anomalous Zeeman effect.

    1923. Adolf Smekal (1895-   ). Prediction of
scattering of radiation by an atom with frequency
modified by a Bohr transition frequency of the atom.

    1923-1924. de Broglie. Waves associated with
matter, interpretation of Bohr quantum condition.

    1924. Wolfgang Pauli (1900-1958). Hyperfine
structure.

    1924. James H. Van Vleck (1899-   ). Correspondence
principle applied to absorption. Classical absorption
corresponds to the difference between absorption and
stimulated emission computed from Einstein's B coefficients. Van Vleck freely uses the phrase "negative
absorption." "The existence of the induced emission term
on the quantum theory may at first sight seem strange,
but it is well known that it is qualitatively explained
in that with the proper phase relations a classical
electric wave may receive energy from an atomic system,
although on the average (i.e., integrating over all
possible phase relations) it contributes more than it
receives in exchange. It is therefore the excess of
positive absorption over the induced emission which one
must expect to find asymptotically (for large quantum
numbers) connected to the net absorption in the classical
theory."

    1924. Satyandra N. Bose (1894-   ). Derived
Planck law from assumption of Bose-Einstein statistics.

    1924. de Broglie. Wave particle duality valid for
particles as well as for light.

    1924. Bohr, Kramers and John C. Slater (1900-   ).
The discovery of the Compton effect made great difficulties for the wave theory, just as diffraction
phenomena did for the light quantum theory. These authors
introduced a notion of "virtual radiation" produced by
an atom in a stationary state, which had all the frequencies which the atom could emit in a real transition.
This radiation was itself not directly observable, but
was able to induce in other atoms the processes considered by Einstein, as well as to induce spontaneous
emission in the first atom itself. The idea of virtual
radiation does to some extent anticipate aspects of the
quantum theory of radiation. A transition of an atom

from one state to another was alleged to be accompanied
by changes in energy and momentum but not by the emission of radiation. Detailed conservation laws of energy
and momentum were given up, but these retained validity
on an averaged basis. The whole discussion was completely qualitative, and fortunately fell quickly out of
favor after the observations of Bothe and Geiger.

        1924. Bothe and Geiger. Disproved speculations of
Bohr, Kramers and Slater.

        1924. Richard C. Tolman (1881-1948). In a paper
"Duration of molecules in upper quantum states", Phys.
Rev. 23, 693 (1924), Tolman stated that "molecules in
the upper quantum state may return to the lower quantum
state in such a way as to reinforce the primary beam
by "negative absorption." Tolman inferred "by analogy
with classical mechanics" that the process of negative
absorption "would presumably be of such a nature as to
reinforce the primary beam." After thus clearly preparing a basis for the invention of the laser, Tolman
pointed out that for absorption experiments as usually
performed the amount of negative absorption can be
neglected.

        1924. Kramers. Generalized Ladenberg dispersion
formula to apply to atoms in excited states. New terms
in the formula had reversed sign leading to "negative
dispersion" corresponding to the "negative absorption"
implied by Einstein's stimulated emission process.

        1924-1925. Einstein. Bose-Einstein statistics
applied to a monatomic gas.

        1925. Pauli. Electron has a two-valuedness not
describable classically, i.e., spin degree of freedom.
Exclusion principle. Hyperfine Structure caused by
nuclear spin.

        1925. Kramers and Werner Heisenberg (1901-   ).
Generalized the radiative scattering theory to be compatible with correspondence principle.

        1925. George E. Uhlenbeck (1900-   ) and Samuel
A. Goudsmit (1902-   ). Electron spin and magnetic
moment (ideas held previously by Compton and R. deL.
Kronig.)

        1925. Walter Elsasser (1904-   ). Suggested dif-
fraction of matter waves to explain 1923 scattering
experiments of Clinton J. Davisson (1881-1958) and
Charles H. Kunsman (1890-   ); and the Ramsauer effect
in scattering of slow electrons by atoms. Electron diffraction was more clearly established in later work of
Davisson and Lester H. Germer (1896-1972), as well as of
George P. Thomson (1892-   ).

        1925. Enrico Fermi (1901-1954) and Franco D.
Rasetti (1901-   ). Applied (non-resonant) radio-

frequency fields to excited atoms.

1925. Compton and Alfred W. Simon (1897-   ).
Disproof of speculations by Bohr, Kramers and Slater.

1925-1926. Heisenberg, Max Born (1882-1970), and
Pascual Jordan (1902-   ). Matrix mechanics.

1925-26. Paul A. Dirac (1902-   ). Operator
formulation of quantum mechanics.

1926. Erwin Schrödinger (1887-1961). Wave Mechanics.

1926. Born. Probability interpretation of the wave
function. Einstein and others had the idea that the
classical intensity of an electromagnetic field was
proportional to the probability of the occurrence of
light quanta.

1926. Gregor Wentzel (1898-   ). Derived Rutherford scattering from Born's probability interpretation
of wave mechanics.

1926. Wentzel. Gave wave mechanical derivation of
probability of photoelectric effect caused by classical
electromagnetic field. See Lamb and Scully in Kastler
Jubilee volume.

1926. Dirac. Time dependent pertubation theory.

1926. Schrödinger. Wave packet for a simple harmonic oscillator.

1927. John B. Johnson (1887-   ). Observed thermal
noise in a resistor.

1927. Earle H. Kennard (1885-   ). Wave packets
for simple harmonic oscillator as well as for free
particle and charge in uniform magnetic field.

1927. Dirac. Quantum theory of radiation.

1927. Kramers and Ralph deL. Kronig (1904-   ),
independently. Derived relation between real and imaginary parts of the index of refraction (or electric susceptibility) of a dielectric medium. Derivation based
on general assumptions that the complex index of refraction is an analytic function of a complex variable
frequency, and that the problem obeys a casuality
condition. The proof uses Cauchy's theorem in the
theory of functions of a complex variable.

1928. Harry Nyquist (1889-   ). Theory of Johnson noise.

1928. Chandrasakara V. Raman (1888-   ). Observation of inelastic scattering predicted by Einstein
and by Smekal.

1928. Ladenburg and Hans Kopfermann (1895-1963).
Confirmed negative dispersion terms by study of refractive index in neon gas excited by an electrical discharge.

1929. Oskar Klein (1894-   ), and Yoshio Nishina
(1890-1951). Theory giving Compton relations and cross
section for Compton effect according to relativistic
quantum mechanics of Dirac.

1930. Heisenberg and Pauli. Quantum theory of radiation, more general than Dirac's treatment.

1930. J. Robert Oppenheimer (1904-1967). First (incomplete) calculation of finite quantum electrodynamic level shift.

1930. Frits Zernike (1888-1966). Phase contrast microscope.

1932. Ivar Waller (1898-   ). Theory of para-magnetic relaxation. Important for nuclear resonance, and solid state lasers.

1932. Fermi. Lectures given at University of Michigan Summer Session of 1930, published in Rev. Mod. Phys. $\underline{4}$, 87 (1932). A very instructive account of the subject, with detailed treatment of the emission of light from an excited atom, propagation of light in vacuum, theory of Lippman fringes and the theory of the Doppler effect, etc.

1932-1933. Gregory Breit (1899-   ). Articles on the quantum theory of dispersion (Rev. Mod. Phys. $\underline{4}$, 504 (1932), $\underline{5}$, 91 (1933). The first of these articles is preceded by an account of experimental work written by Korff and Breit.

1933. Otto R. Frisch (1904-   ). Experimental confirmation of momentum transfer to an atom emitting or absorbing radiation.

1933. Pauli. Articles on the principles of quantum mechanics in Handbuch der Physik.

1933. Bohr and Leon Rosenfeld (1904-   ). Theo-retical discussion of measurement of electromagnetic fields in quantum electrodynamics.

1933-1935. Frederick B. Llewellyn (1897-1971). Diode electron-tube transit time oscillator.

1934. Claud E. Cleeton (1907-   ) and Neil H. Williams (1870-1956). Microwave spectroscopy of ammonia molecule using magnetron oscillator of wave-length 1.25 cm.

1934-1937. Pavel A Cerenkov (1904-   ), Il'ja M. Frank (1908-   ), and Igor E. Tamm (1895-1971). Dis-covery and theory of the Cerenkov effect.

1935. F. M. Penning (1894-1953), et al. Photo-electric cell with single stage secondary emission amplification, followed a year later by Vladimir K. Zworykin (1889-   ), et al with multi-stage photo-multiplier tube.

1936. Walter H. Heitler (1904-   ). Monograph Quantum theory of radiation. Heitler gave a detailed calculation showing how a simple harmonic oscillator could lose energy when acted on by a periodic driving force. He regarded this as a classical counterpart of induced emission. However, when he averaged the power

transfer over all phases, he found only a net absorption
of energy by the oscillator systems.

    1937. I. I. Rabi (1898-   ) and Julian S. Schwinger
(1918-   ). Magnetic resonance of spins in an atomic
beam.

    1938. Felix Bloch (1905-   ). Magnetic resonance
of neutron spins, used in 1939 experiments with Luis
W. Alvarez.

    1939. Russell H. Varian (1898-1959) and Sigurd F. Varian
(1901-1961). Invention of the klystron microwave generator.

    1945. Evgenii K. Zavoisky (1907-   ). Para-
magnetic resonance in solids.

    1946. Edward M. Purcell (1912-   ), Henry C. Torrey
(1911-   ) and Robert V. Pound (1919-   ). Observed
nuclear magnetic resonance absorption. Independently
discovered by

    1946. Bloch, William W. Hansen (1909-1949), and
Martin E. Packard (1921-   ). Nuclear induction.

    1947. Willis E. Lamb, Jr. (1913-   ) and Robert
C. Retherford (1912-   ). Fine structure of hydrogen
by a microwave method. Appendix I of a paper on this
work (Phys. Rev. _79_, 549 (1950) is entitled "Conditions
in a Wood's Discharge." This represents a missed opportunity to invent the maser, and hence it may be of
interest to give the text here.

"The absorption of radio waves by excited hydrogen
atoms in a Wood's discharge tube depends on the populations of the various states. These in turn depend on
the rates of production and decay of the excited atoms.
It would involve a lengthy program of research to make
quantitative calculations of these, and we shall be
content here with the roughest sort of estimate. Only
the n=1 and n=2 states of atomic hydrogen will be considered.

We shall assume that the 2p states decay to 1s at
a rate corresponding to the natural lifetime $\tau_p = 1.6 \times 10^{-9}$
sec., and that the excitation of 2p is due primarily to
two causes: (1) electron bombardment and (2) absorption
of the Lyman resonance radiation emitted by other atoms.
Since the absorption coefficient for this radiation is
very large, a resonance radiation quantum can, on the
average, only escape from the tube after a large number
of absorptions and re-emissions. This number has been
estimated to be 500 to 1000 for typical discharge conditions. As a result, the effective decay rate of the
2p states is much smaller than that given by the natural
lifetime, and the population of states 2p is correspondingly increased.

The situation is markedly different for the $2^2S_{1/2}$ state. As indicated in Section 9, excitation from $1^2S_{1/2}$ electron bombardment is about one-tenth as likely as for 2p excitation. The imprisonment of resonance radiation plays no role here, for the $2^2S_{1/2}$ state cannot combine optically with the ground state. As a result, however, the $2^2S_{1/2}$ state under certain circumstances may be metastable, and despite the low rate of production, the population increased. As shown in Section 6, the decay rate of $2^2S_{1/2}$ increases with the electric field acting on the atom, and decreases with increasing separation of the states $2^2S_{1/2}$ and $2^2P_{1/2}$. If this separation is zero as given by the Dirac theory, the life of $2^2S_{1/2}$ in a typical Woods discharge is a few times the natural life, but if the separation corresponds to 1000 Mc/sec., the life is increased to about 900 τ_p, and the population of $2^2S_{1/2}$ is correspondingly increased.

Let us consider the transitions between $2^2S_{1/2}$ and $s^2P_{3/2}$ induced by radio waves. If these states are populated in accordance with their statistical weights (equipartition), there will be no appreciable net absorption of r-f since the induced emission exactly cancels the induced absorption. (Spontaneous transitions between $2^2P_{3/2}$ and $2^2S_{1/2}$ occur at a negligible rate.) If the population of $2^2S_{1/2}$ is increased relative to $2^2P_{3/2}$, there will be a net absorption of r-f. If, on the other hand, $2^2P_{3/2}$ is more highly populated, there will be a net induced emission (negative absorption!)

On the basis of the preceding discussion alone, one would expect that the 2p levels would be about five to ten times more populated than the 2s levels. In that case, one would expect to find a negative absorption,

and as estimated below, a large one. Admittedly the
population estimates are rough. Still it would seem to
require an accident for exact equipartition, unless there
is some mechanism efficiently coupling $2^2S_{1/2}$ and $2^2P_{1/2}$.
As indicated above, and as described quantitatively in
Section 6, there is Stark effect coupling between these
two states. If it were not for the entrapment of
resonance radiation, the $2^2P_{1/2}$ state would not be sufficiently populated for the establishment of equipartition by this mechanism, but on the basis of the
numerical estimates given above, and in Appendix II, it
seems quite possible that it can occur, and that the
absorption of r-f would be markedly reduced.

   If equipartition is not achieved, a net absorption
or induced emission can occur. In order to form an
idea of the expected magnitude, we shall calculate the
unbalanced rate of transitions induced by r-f from $2^2S_{1/2}$
to $2^2P_{3/2}$ ignoring the reverse transitions which might
nearly cancel out or reverse the sign of the whole
effect. Although the result may be an overestimate of
the expected "absorption", it provides a convenient
basis for discussion.

   We proceed to estimate this absorption rate. In
a typical Wood's discharge tube, there might be a pressure of 0.15 mm of mercury corresponding to a density of

$$n_H = (0.15 \times 2.687 \times 10^{19})/(760) = 5.3 \times 10^{15}$$

hydrogen atoms per cubic centimeter. The number of
atoms excited to $2^2S_{1/2}$ per unit volume per second is
given by

$$J\sigma n_H$$

where eJ is the electron current density, and σ is the
excitation cross section. If we equate this to the assumed rate of density decay

$$n*/(900\tau_p)$$

we obtain

$$n* = 900 \ J\sigma\tau_p n_H \ .$$

Taking an electron current density corresponding to 0.1 amp./cm^2

$$J=(0.1)/(1.602 \times 10^{-19}) = 6.24 \times 10^{17} \quad \text{electron/cm}^2/\text{sec.}$$

and

$$\sigma = 10^{-17} \ \text{cm}^2$$

(Section 9), we find

$$n* = 4.7 \times 10^{10} \ \text{cm}^{-3}$$

in state $2^2S_{1/2}$.

We now consider the absorption of radio waves by these excited hydrogen atoms, taking as explained above, only the transitions to $2^2P_{3/2}$ into account. As in Section 13, the transition probability for absorption of radiation is

$$1/\tau \ \text{induced} = (2\pi e^2 S_0 \gamma/ch^2)|(|\hat{e}. \ \vec{r}|)|^2 \ /$$
$$[(\omega-\omega_0)^2 + (\gamma/2)^2],$$

where S_0 is the incident energy flux density of the radiation having circular frequency ω and electric polarization vector parallel to the unit vector $\hat{e}$, ω_0 is the circular resonance frequency, $(|\vec{r}|)$ is the matrix element of the coordinate vector r of the atomic electron for the transitions from $2^2S_{1/2}$ to $2^2P_{3/2}$, $\gamma = 1/\tau_p$ is the reciprocal lifetime of the $2^2P_{3/2}$ state. For some estimates it is useful to use instead of a transition probability a cross section σ_{induced} for the absorption of a microwave photon. This is given by the equation

$$J_p \sigma_{\text{induced}} = 1/\tau \ \text{induced}$$

where J_p is the flux density of such photons. Now $J_p = S_0/\hbar\omega_0$ hence

$$\sigma_{\text{induced}} = \hbar\omega_0/S_0\tau$$

or

$$\sigma_{induced} = 2\pi(e^2/\hbar c)|(|\hat{e}\cdot\vec{r}|)|^2 \omega_0 \gamma/[(\omega-\omega_0)^2+(\gamma/2)^2].$$

For the sum of all transitions to the sub-levels of $2^2P_{3/2}$ one has $|(|\hat{e}\cdot\vec{r}|)|^2$ equal to two-thirds of the value in which the sum is taken over all sub-levels of $2^2P_{1/2}$ and $2^2P_{3/2}$. This sum is just the value which would be calculated for the transitions from 2s to 2p ignoring electron spin. Hence

$$|(|\hat{e}\cdot\vec{r}|)|^2 \rightarrow (\tfrac{2}{3})(27a_0^2/3) = 6a_0^2 .$$

At resonance

$$\omega = \omega_0 = 2\pi(10,950) \times 10^6 \text{ sec.}^{-1}$$

and using

$$\gamma = 1/(1.595 \times 10^{-9}) = 6.25 \times 10^8 \text{ sec.}^{-1}$$

we find

$$\sigma_{induced} = 3.4 \times 10^{-15} \text{ cm}^2 .$$

For $n^* = 4.7 \times 10^{10}$ atoms/cm^3 in state $2^2S_{1/2}$, the absorption coefficient would be $\mu = 1.6 \times 10^{-4}$ cm^{-1}. This
is a large absorption coefficient by modern microwave
spectroscopic standards, but the breadth of the resonance $\gamma/2\pi \sim 100$ Mc/sec. is many times the widths
usually met in that field. As explained above, it is
possible that this absorption is nearly all canceled
out by the inverse transitions. In view, however, of
the extreme crudeness of the numerical estimates, it is
possible that some appreciable departure from equipartition may exist, and that an absorption or induced
emission could be detected. It is therefore highly
desirable that a search for such effects should be made,
especially under discharge conditions which do not favor
equipartition.

As a complicating factor, there will be a large
and frequency dependent background absorption of microwaves in the discharge due to the electrons. Haase
found good agreement with an equation derived by Stewart

$$\mu_{electrons} = (2e^2 z \lambda^2 N_e)/(\pi m c^3)$$

for the absorption coefficient of radiation with wavelength λ. Here z is the number of collisions with gas
molecules per second experienced by an electron, and
N_e is the number of free electrons per cubic centimeter.
For the typical figures

$$z = 0.95 \times 10^9 \text{ sec.}^{-1}$$
$$N_e = 2 \times 10^9 \text{ cm}^{-3} \quad .$$

$$\lambda = 3 \text{ cm}$$
$$\mu_{electrons} = 1.03 \times 10^{-4} \text{ cm}^{-1} \quad . \quad "$$

     The concept of negative absorption was new to us at
the time, and we were unaware of the earlier references
to it given in the Chronological Listing. I think
that we understood that the radiation would be coherent,
as was the input signal. However, we did not associate
negative absorption with self-sustained oscillation.
Even if we had done so, at least three factors would have
kept us from inventing the maser: (1) our interest was
centered on the fine structure of hydrogen, (2) the
smallness of the expected absorption (gain), and doubt
as to its sign, and (3) the ready availability of
oscillators at the frequency used.
     1947. Denis Gabor (1900-   ), First suggestion
of holography. Full realization only after invention
of the laser.
     1948. Van Vleck. More detailed theory of para-
magnetic resonance.
     1948. Robert Karplus (1927-   ), and Schwinger.
Theory of saturation resonance broadening.
     1948-1949. Sommerfeld. "It is strange that practical
electronics remained untouched by these fundamental
facts (i.e., quantum theory, electron spin and magnetic
moment) and could get along with the notion of the charged
point mass or the minute charged sphere." (From Electrodynamics, p. 279.) On page 236, Sommerfeld attributes
to Einstein the saying "The electron is a stranger in
electrodynamics." Some quotations from Optics, Page 1,
are: "Leonardo da Vinci called optics the paradise of
mathematicians. Of course, by optics he meant only
geometrical or ray optics, the theory of perspective and
the distribution of light and shadow," and "It was the
tragedy in the life of Goethe that he would not recognize
the distinction between physical and physiological optics:
this was the reason for his fruitless fight against

Newton." On Page 97, Sommerfeld recalls a letter from
Drude early in this century which states "We live in a
grandiose era; we are beginning to get a glimpse of the
electric composition of matter." See also the Chronological Listing for 1900.

   1948-1949. William Shockley (1910-   ), John
Bardeen (1908-   ) and Walter H. Brattain (1908-   ).
Invention of the transistor.

   1949. Francis Bitter (1902-1967). Optical detection of magnetic resonance. Analysis criticized by
Maurice H. Pryce.

   1949. Jacques Brossel and Alfred Kastler (1902-   ).
Double resonance.

   1950. Alfred Kastler. Suggested method of optical
pumping.

   1951. Harold I. Ewen and Purcell. Hyperfine
structure of hydrogen atoms in our galaxy.

   1951. Charles H. Townes (1915-   ). Concept of
maser. This idea was worked on with a post-doctoral
fellow, H. J. Zeiger, and a graduate student, J. P.
Gordon. The research was carried out in the Columbia
Radiation Laboratory at Columbia University and support-
ed by the U. S. Army Signal Corps and the Office of
Naval Research. The quarterly reports of the Labora-
tory's work were given a very wide circulation among
scientists known to be interested in microwave physics.
The first mention of the maser project appears in the
December 31, 1951, quarterly report under the names of
Zeiger and Gordon. (Some, but not all, of the research
supervisors modestly did not put their names on reports.)
The text follows:

   "8. Molecular Beam Oscillator (H. J. Zeiger and J.
P. Gordon).

   Preliminary calculations have been performed
on the design of a molecular beam oscillator. The essential elements of the oscillator are:
   1. A molecular beam source.
   2. A deflecting region for separating an excited
state of the beam from some ground state.
   3. A cavity, tuned to the frequency of transition
from the excited state to the ground state by induced
emission.
   4. A detector for observing the radiation from the
transitions.

   The projected oscillator is intended for use in the
long wavelength infrared ($\lambda \sim$ .5mm). To be suitable

as a source of radiation, a molecule must have large
rotational constants, and a large electric dipole moment.
A convenient molecule for consideration is ND_3 which has
B_e = 5.18 cm^{-1}, C_e = 3.15 cm^{-1}, and μ = 1.57 Debye.

The amount of molecular beam available for use in
the oscillator is limited because the pressure behind
the source slit must be low enough to give effusive
flow from the slit. For a slit of dimension 2cm by 1
mil, a resonable pressure is 2 mm of Hg. In this case,
the total number of molecules per second effusing from
the slit would be 5.3 × 10^{18} for ND_3 at dry ice temperature.

The separation of the excited state from a ground
state in the deflecting region is based on the linear
Stark effect in a symmetric top molecule. The Stark
energy, and therefore the effective electric dipole
moment of a molecule in an electric field depends on the
quantum numbers J, K, and M of the state. In a Stern-
Gerlach inhomogeneous electric field, a separation of
states is effected. However, because the length of the
path must be about 10 cm for good separation, the solid
angle of beam subtended at the cavity entrance would be
small ($\sim$ 3 × 10^{-5}).

To increase the effective solid angle of the beam
at the cavity, a focuser of the type described by Friedberg and Paul (Friedberg and Paul, Naturwiss., 38, Heft
7, 1951.) was considered. In this type of focuser,
with two dimensional equipotentials of the form $V=Ar^3$.
cos 30, a Hooke's law radial restoring force acts on a
molecule with a linear Stark effect. The focuser not
only increases the effective solid angle of the beam,
but effects a partial selection of the upper state,
because the focusing depends on effective dipole moment.

To conform to the radial symmetry of focusing, the
source slit was taken to be annular, of dimension 3 mm
in radius by 1 mil in width. The cavity was taken as
cylindrical, with an entrance slit 3 mm in radius by
20 mils in width. The transition considered in ND_3 was
(J=2, K=1, M=2 → J=1, K=1, M=1), which has a transition
energy of 20.55 cm^{-1}. With a voltage of 10,000 V.
across the plates, and a gap of 1 cm diameter, the focuser
must be 35.6 cm long to focus the upper state (J=2, K=1,
M=2). The effective solid angle of beam passing through
the focuser would then be $\sim$ 4.7 × 10^{-3}.

A calculation by George Dousmanis shows that 3/10 of the total beam in the upper state, passing through the focuser, would enter the annular slit in the cavity. The population of this state at dry ice temperature is 0.24%. Then, the total amount of beam in the upper state entering the cavity would be ~6 × 10^{12} molecules/sec. If all of these were to undergo induced transitions to the ground state, ~2.4 × 10^{-9} watts of power would be delivered to the cavity. This is enough power to detect with a Golay cell.

A theoretical calculation of the Q of a tuned cylindrical cavity was performed. For a cavity 1 cm in diameter and 1 cm in length, at liquid air temperatures, and tuned to 20.55 cm^{-1}, the calculated Q is 1.5 × 10^{5}. The Q calculated as necessary to maintain oscillations for 2.4 × 10^{-9} watts input is ~1 × 10^{5}. On the basis of these calculations, it therefore seems only barely possible that oscillations will be sustained in the cavity."

In the next Quarterly Report, the proposed goal was changed to K-band operation using the ammonia J=3, K=3 inversion transition at 1.25 cm wavelength. Subsequent reports told details of the vacuum system, focusser, cavity resonator and microwave resonator. Operation as an oscillator was first mentioned on January 30, 1954, in much the same form as in the published letter of Phys. Rev. 95, 282 (1954).

It is interesting to note that nowhere in these reports is there mention of coherence of the radiation. It is possible that these features were too obvious for mention, but in view of the very fuzzy ideas of stimulated emission in the history of the subject, it was certainly not clear that the radiation generated would be coherent.

The only published paper on this work prior to its success was by C. H. Townes, J. Inst. Elec. Commun. Eng. (Japan) 36, 650 (1953). The term "maser" was first used in the full length paper of Phys. Rev. 99, 1264 (1955).

1951. Walter Gordy (1909-   ). Monograph Microwave Spectroscopy.
1951. Purcell and Pound. Population inversion in lithium fluoride. Concept of negative spin temperature.

1953. Joseph Weber (1919-   ). Paper "Amplification of microwave radiation by substances not in
thermal equilibrium", Trans. Inst. Radio Engrs. Professional Group on Electron Devices, 3 June, 1953. Based
on talk given at the 1952 Ottawa electron tube conference.
The proposed methods for achieving population inversion
have not been reduced to practice, and do not seem
very practicable. Although the word coherent is used,
only an amplifier is under consideration. Nothing is
said of self-sustained oscillation.

1954. Gordon, Zeiger and Townes. First maser
amplifier, spectrometer and oscillator in operation.

1954. Robert H. Dicke (1916-   ). Paper "Coherence in Spontaneous Radiation Processes." Concept
of super-radiance. See also "The Coherence Brightened
Laser," Ref. and Dicke's Patent "Molecular Amplification and Generation Systems and Methods" (application
1956, Granted, 1958) as reprinted in Reference 71.

1954. Nicholas Bloembergen (1920-   ) and Pound.
Paper "Radiation Damping in Magnetic Resonance Experiments." Coupled Kirchoff circuit and Block NMR equations.
Selfconsistency method. Transient solutions. No theory
of CW oscillator.

1954-1955. Israel R. Senitzky (1920-   ). Use
of Shrödinger wave packet to describe coherent electromagnetic field.

1954. Nikolai G. Basov (1922-   ) and Alexander
M. Prochorov (1916-   ). Idea for a gas maser.

1955. Arthur T. Forrester (1918-   ), et al.
Photoelectric mixing of incoherent light. Experimental evidence that photoemission is proportional to
the square of the total wave amplitude incident on the
photosurface.

1955. Townes and Arthur L. Schawlow (1921-   ).
Monograph Microwave Spectroscopy.

1955. Basov and Prokhorov. Idea for a three-level
solid state maser.

1956. R. Hanbury Brown (1916-   ) and R. Q. Twiss.
Correlations in photoelectric currents for cells illuminated by two light beams.

1956. Bloembergen. Proposal for a three level
solid state maser.

1956. John C. Helmer (1926-   ) and Lamb. Semi-
classical theory of ammonia beam maser. See Stanford
University Microwave Laboratory Technical Report No.
ML-311 (1956) by J. C. Helmer. Also Helmer's Stanford
University Thesis "Maser Oscillators" (University Micro-
films) and Lamb "Quantum Mechanical Amplifiers" in
Lectures in Theoretical Physics, Vol. II, edited by W.E.
Brittin and B.W. Downs, (Interscience, New York, 1960).

Detailed calculations of operating conditions of NH_3 maser for Maxwellian distirbution of velocities. Townes (J. Appl. Phys. 28, 920 (1957)) showed that one of the predictions failed because of non-Maxwellian velocity distribution of beam molecules caused by collisions.

Anatole Abragam (1914-   ), et al. Proton maser oscillator operating in Earth's field.

1957. Richard P. Feynman (1918-   ), Frank L. Vernon, and Robert W. Hellwarth (1930-   ). Schematic formulation of maser theory, with emphasis on geometrical representation of two level atomic system.

1957. Philip W. Anderson (1923-   ). Paper "The reaction field and its use in some solid state amplifiers" J. Appl. Phys. 28, 1049 (1957). Self-consistency semi-classical treatment of maser problems.

1957. Henry E. Scovil (1923-   ), George Feher (1924-   ) and H. Seidel. Successful operation of a three level solid state maser of the type proposed by Bloembergen in 1956.

1957-1958. Gordon Gould (1920-   ). Suggested Fabry-Perot cavity and use of alkali vapor, ruby and rare earth crystals as working substance. He also considered laser action in an electrical discharge, and enhancement of population inversion by collisions depleting lower states. Part of a classified proposal and never published. An account of the unsuccessful patent suit brought against the Townes patents is given by Max Gunther, "Laser: The Light Fantastic" in Playboy Magazine, February, 1968.

1958. G. Makhov, C. Kikuchi, J. Lambe and R. W. Terhune. Maser action in ruby.

1958. Townes and Schawlow. Published "Infrared and Optical Masers", Phys. Rev. 112, 1940 (1958), with detailed proposal for an optical maser based on potassium vapor. Fabry-Perot resonator.

1958. Ali Javan (1926-   ). Proposed amplification by Raman processes, without need for population inversion.

1959. William Culshaw (1914-   ). Important studies of properties of ruby relevant for optical maser action.

1959. Javan. Paper "Possibility of production of negative temperatures in gas discharges," in Phys. Rev. Letters 3, 87 (1959). Concrete proposal of first gas laser.

1959. John H. Sandars (1924-   ). Suggested a mechanism for laser action.

1959. F. A. Butayeva and V. A. Fabrikant, of the Institute of Optics of the U.S.S.R. Academy of Sciences wrote "A medium with negative absorption" in a memorial volume to G. S. Landsberg: Investigations in Experi-

mental and Theoretical Physics, U.S.S.R. Acad. Sci.
Publisher, 1959. This work, done in 1957, claimed
amplification of light in a mercury vapor-hydrogen
discharge, in which excitation transfer produced population inversion and negative absorption.

According to Carrol (Ref. 76), there is a Russian
patent now dated 1951, but not published until 1959,
awarded to V. A. Fabrikant of the Moscow Power Institute.
(Apparently the Soviet Patent Office at first rejected
the claim.) The patent reads:

"Methods of amplifying electromagnetic radiation
(ultraviolet, visible, infrared, and radio wave bands)
distinguished by the fact that the amplified radiation
is passed through a medium which, by means of auxiliary
radiation, or by other means, generates excess concentration, of atoms, other particles, or systems at upper
energy levels corresponding to excited states."

Garrett (Ref. 69) has even traced similar work of
Fabrikant back to 1941. There is no evidence that laser
action evolved from this work. In fact, Sanders, et al
in "Search for light amplification in a mixture of mercury vapor and hydrogen", Nature 193, 767 (1962) were
unable to repeat the work of Butayeva and Fabrikant.
It is possible that population inversion was achieved
by V. K. Abelkov, et al in J.E.T.P. 12, 618 (1960) and
by Ferkhman and Frish, p. 1936.
    1960. Theodore H. Maiman (1927-   ). First
optical maser, in ruby. Phys. Rev. Letters 4, 564 (1960).
    1960. Arthur G. Fox (1912-   ) and Tingye Li
(1931-   ). Theory of resonant modes in a Fabry-Perot
interferometer.
    1961. Javan, William R. Bennett, Jr. (1930-   ),
and Donald R. Herriott (1928-   ). Article "Population
Inversion and Continuous Optical Maser Oscillations in
a Gas Discharge Containing a He-Ne Mixture," (Phys.
Rev. Letters 6, 106, (1961)).
    1961. Peter A. Franken (1928-   ), et al.
Generation of optical harmonics. Also optical mixing
and rectification. Laid foundation for non-linear optics,
to which major contributions were made by Bloembergen,
Terhune, Giordmaine, etc.
    1961. Ramsey, et al. Hydrogen h.f.s. maser.
    1962. R. N. Hall, et al; M. I. Nathan, et al;and
T. M. Quist, et al. Semiconductor lasers.
    1962. Eric J. Woodbury (1925-   ) and W. K. Ng.
Stimulated Raman emission. Theory by Hellwarth.
    1963. Roy J. Glauber (1925-   ). Quantum theory
of optical coherence.

1963. Ross A. McFarlane (1931-   ), Bennett and
Lamb. Confirmation of central tuning dip in output
of a gas laser predicted by semi-classical theory of
an optical maser.

1964. F. Schwabel and Walter E. Thirring (1927-   ).
In article "Quantum Theory of Laser Radiation" (Erg. d.
exakten Naturw. 36, 219 (1964)) attempted to treat the
active medium as made of atoms with repulsive Hooke's
law forces.

1964. Lamb. "Theory of an Optical Maser," in Phys.
Rev. 134, A1429 (1964). Discussed realistic model for
active medium. Predicted levels and frequencies of
operation of multi-mode operation. The prediction of the
central tuning dip was communicated by letters to Javan
and Bennett in 1962. The theory was described at the
Third International Laser Conference in Paris, February,
1963.

1965. H. Weaver, D. R. Williams, et al. Discovery
of cosmic maser amplification of microwave radiation
from the hydroxyl free radical.

1966-1967. Marlan O. Scully (1939-   ) and Lamb.
"Quantum Theory of an Optical Maser," Phys. Rev. Let-
ters, 16 853 (1966) and Phys. Rev. 159, 208 (1967).
Fully quantum mechanical theory of a laser.

1967. R. L. Pfleegor and Mandel. Interference produced in beams from two lasers at low light levels.

1969. Gerald Feinberg (1933-   ). In Scientific
American reprint collection "Lasers and Light," states
"What the laser does is to produce vast numbers of particles of exactly the same energy and wavelength. With
no other stable particle but the photon is such a feat
possible. The laser beam's remarkable macroscopic
properties arise from the fact that its constituent
photons are precisely identical. Whether the laser
could have been invented without quantum mechanics is
an interesting question." And "At present the photon
theory gives an accurate description of all that we
know about light." Science sometimes has setbacks!

1972. Matthew Borenstein (1943-   ) and Lamb.
Paper "Classical Laser" in Phys. Rev.   (1972). Theory
of laser action using only classical mechanics and
electromagnetic theory.

III. REFERENCE BOOKS

A number of important books have been mentioned in
the Chronological Listing. It seems more convenient
to provide a separate listing of more recent books, of
which only a few are included in the Section II. The
main objective of the following list is convenience,
and no attempt at completeness is made. The books in
each group are arranged approximately in order of publication of the first edition.

History

1. E. T. Whittaker, History of the Theories of
Aether and Electricity. Two volumes. (Nelson, London,
1951-1953).
2. W. F. Magie, A Source Book in Physics. (McGraw-
Hill, New York, 1965).
3. J. Jammer, Concepts of Quantum Mechanics.
(McGraw-Hill, New York, 1966).
4. A. I. Sabra, Theories of Light from Descartes
to Newton. (Oldbourne, London, 1967).
5. D. ter Haar, The Old Quantum Theory. (Pergamon,
Oxford, 1967).
6. B. L. van der Waerden, Sources of Quantum
Mechanics. (Dover, New York, 1967).

Electricity and Magnetism

7. M. Abraham and R. Becker, Theorie der Elek-
trizitat, in many editions, with varying authors,
starting in 1894.
8. J. Jeans, Mathematical Theory of Electricity
and Magnetism. (Cambridge, 1925).
9. H. A. Lorentz, Theory of Electrons. (Dover,
New York, 1952).
10. O. W. Richardson, The Electron Theory of Matter.
(Cambridge, 1914).
11. A. Sommerfeld, Electrodynamics. Vol. III of
Lectures on Theoretical Physics. (Academic Press, New
York, 1952).
12. L. Rosenfeld, Theory of Electrons. (Interscience,
New York, 1951).
13. W. K. Panofsky and M. N. Phillips, Classical
Electricity and Magnetism. (Addison-Wesley, Cambridge,
1955).
14. L. D. Landau and E. M. Lifshitz, Electrodynamics
of Continuous Media. Vol. VIII of Course of Theoretical
Physics. (Pergamon, Oxford, 1960).

15. J. D. Jackson, Classical Electrodynamics. (Wiley, New York, 1962).

Optics

16. F. Maurolicus, Illumination on Light. Translated by H. Crew. (MacMillan, New York, 1940). A good treastise on medieval optics.
17. I. Newton, Opticks. (4th ed., 1730; Dover reprint, 1952).
18. J. W. von Goethe, Theory of Colours, (M.I.T. Press, Cambridge, 1970).
19. T. Preston, The Theory of Light. (MacMillan, London, 1928).
20. P. K. Drude, The Theory of Optics. (Longmans, London, 1902).
21. J. Walker, Analytical Theory of Light. (Cambridge, 1904).
22. R. W. Wood, Physical Optics. (MacMillan, New York, 1934).
23. E. Mach, The Principles of Optics. (Methuen, London, 1926; Dover reprint 1953).
24. M. Born, Optik. (Springer, Berlin, 1933).
25. F. A. Jenkins and H. E. White, Fundamentals of Optics. (McGraw-Hill, New York, 1937).
26. A. Sommerfeld, Optics. Vol. IV of Lectures on Theoretical Physics. (Academic Press, New York, 1952).
27. B. Rossi, Optics. (Addison-Wesley, Reading, Mass., 1957).
28. M. Born and E. Wolf, Principles of Optics. (Pergamon, New York, 1964).

Spectroscopy (at all frequencies)

29. H. E. White, Introduction to Atomic Spectra. (McGraw-Hill, New York, 1934).
30. E. U. Condon and G. H. Shortley, The Theory of Atomic Spectra. (Cambridge Press, 1935).
31. A. H. Compton and S. K. Allison, X-rays in Theory and Experiment. (Van Nostrand, New York, 1935).
32. G. Herzberg, Atomic Spectra and Atomic Structure. (Prentice-Hall, 1937; Dover reprint).
33. G. Herzberg, Infrared and Raman Spectra of Polyatomic Molecules. (Van Nostrand, New York, 1945).
34. G. Herzberg, Molecular Spectra and Molecular Structure. (Van Nostrand, New York, 1950).
35. W. Gordy, et al., Microwave Spectroscopy. (Wiley, 1953).
36. C. H. Townes and A. L. Schawlow, Microwave Spectroscopy. (McGraw-Hill, New York 1955).

37. N. F. Ramsey, Molecular Beams. (Oxford, 1956).

38. R. A. Smith, et al., The Detection and Measurement of Infra-red Radiation. (Clarendon Press, Oxford, 1957).

39. A. Abragam, The Principles of Nuclear Magnetism. (Oxford, 1961).

Miscellaneous

40. J. W. S. Rayleigh, The Theory of Sound, (MacMillan, London, 1894-1896).

41. G. Duffing, Erzwungene Schwingungen bei Veränderlicher Eigenfrequenz. (Braunschweig, 1918).

42. P. Debye, Polar Molecules. (Chemical Catalogue, New York, 1929; Dover reprint 1945).

43. A. L. Hughes and L. A. Du Bridge, Photoelectric Phenomena. (McGraw-Hill, New York, 1932).

44. J. H. Van Vleck, The Theory of Electric and Magnetic Susceptibilities. (Oxford, 1932).

45. L. N. Ridenour, editor, Radiation Laboratory Series., especially Vol. 6, Microwave Magnetrons and Vol. 7, Klystrons and Microwave Triodes. (McGraw-Hill, New York, 1948).

46. W. Shockley, Electrons and Holes in Semiconductors, with Applications to Transistor Electronics. (Van Nostrand, New York, 1950).

47. C. Kittel, Introduction to Solid State Physics. (Wiley, New York, 1953).

48. H. Bruining, Physics and Applications of Secondary Electron Emission. (McGraw-Hill, New York, 1954).

49. D. ter Haar, Elements of Statistical Mechanics. (Rinehart, New York, 1954).

50. L. B. Loeb, Basic Processes of Gaseous Electronics. (Univ. of California Press, 1955).

Quantum Mechanics

51. W. Pauli, Die Allgemeinen Prinzipien der Wellenmechanik. In Handbuch der Physik, Vol. 24/1. (Springer, Berlin, 1933).

52. H. A. Bethe, Quantenmechanik der Ein-und Zwei-Elektron-enprobleme, in Handbuch der Physik, Vol. 24/1. (Springer, Berlin, 1933).

53. W. Heitler, Quantum Theory of Radiation. (Clarendon Press, Oxford, 1936, also 3rd edition, 1953).

54. L. I. Schiff, Quantum Mechanics. (McGraw-Hill, New York, 1949), (also 3rd edition, 1968).

55. J. Schwinger, editor of Selected Papers on Quantum Electrodynamics. (Dover, New York, 1948).

56. G. Baym, <u>Lectures</u> <u>on</u> <u>Quantum</u> <u>Mechanics</u>.
(Benjamin, New York, 1969).

Quantum (and other) Electronics

57. J. C. Slater, <u>Microwave</u> <u>Electronics</u>. (Van
Nostrand, New York, 1950).
58. C. H. Townes, editor, <u>Quantum</u> <u>Electronics</u>, <u>A</u>
<u>Symposium</u>. (Proceedings of what was to become the first
international conference on quantum electronics) (Held
at High View, N.Y. in September, 1959), (Columbia,
New York, 1960).
59. A. A. Vuylsteke, <u>Elements</u> <u>of</u> <u>Maser</u> <u>Theory</u>.
(Van Nostrand, Princeton, New Jersey, 1960).
60. J. Singer, editor, Advances In Quantum Electronics. (Proceedings of the second international
conference on quantum electronics, held at Berkeley,
California in March, 1961) (Columbia, New York, 1961).
61. J. Fox, editor. <u>Optical</u> <u>Masers</u>. (Polytechnic
Press, Brooklyn, 1963).
62. W. H. Louisell, <u>Radiation</u> <u>and</u> <u>Noise</u> <u>In</u> <u>Quantum</u>
<u>Electronics</u>. (McGraw-Hill, New York, 1964).
63. P. A. Miles, editor, <u>Proceedings</u> <u>of</u> <u>International</u>
<u>School</u> <u>of</u> <u>Physics</u> <u>'Enrico Fermi'</u>. Course XXXI. <u>Quantum</u>
<u>Electronics</u> <u>and</u> <u>Coherent</u> <u>Light</u>. (Academic Press, New
York, 1964).
64. P. Grivet and N. Bloembergen, editors, <u>Quantum</u>
<u>Electronics</u>: <u>Proceedings</u> <u>of</u> <u>Third</u> <u>International</u> <u>Conference</u>
<u>on</u> <u>Quantum</u> <u>Electronics</u>, February, 1963. (Columbia, New
York, 1964).
65. N. Bloembergen, <u>Nonlinear</u> <u>Optics</u>. (Benjamin,
New York, 1965).
66. <u>Masers</u> <u>and</u> <u>Optical</u> <u>Pumping</u>. (American Institute
of Physics, New York, 1965).
67. C. DeWitt, editor, <u>Quantum</u> <u>Optics</u> <u>and</u> <u>Electronics</u>.
Summer School of Physics, Les Houches, 1964. (Gordon and
Breach, New York, 1965).
68. P. L. Kelley, B. Lax and P. E. Tannenwald,
editors, <u>Physics</u> <u>of</u> <u>Quantum</u> <u>Electronics</u> <u>Conference</u>
<u>Proceedings</u>, Puerto Rico, June 1965. (McGraw-Hill, New
York, 1966).
69. C. G. Garrett, <u>Gas</u> <u>Lasers</u>. (McGraw-Hill, New
York, 1967).
70. J. Weber, editor, <u>Masers</u>. <u>A</u> <u>Collection</u> <u>of</u> <u>Re-</u>
<u>prints</u> <u>with</u> <u>Commentary</u>. (Gordon and Breach, New York,
1967).
71. J. Weber, editor, <u>Lasers</u>. <u>A</u> <u>Collection</u> <u>of</u> <u>Re-</u>
<u>prints</u> <u>with</u> <u>Commentary</u>. Two volumes. (Gordon and Breach,
New York, 1968).
72. G. C. Baldwin, <u>Introduction</u> <u>to</u> <u>Nonlinear</u> <u>Optics</u>.

(Plenum, New York, 1969).

73. Polarization, Matière et Rayonnement. (Jubilee Volume in honor of Alfred Kastler.) (Presses Universitaires de France, Paris, 1969).

74. A. Maitland and M. H. Dunn, Laser Physics. (North-Holland, Amsterdam, 1969).

75. A. L. Schawlow, editor, Lasers and Light. Readings from the Scientific American (Freeman, San Francisco, 1969). See especially the following articles:

Feinberg	Light
Weisskopf	How light interacts with matter
Javan	The optical properties of materials
Bloom	Optical pumping
Jones	How images are detected
Gordon	The maser
Schawlow	Optical masers
Schawlow	Advances in optical masers
Pimentel	Chemical lasers
Lempicki and Samuelson	Liquid Lasers
Patel	High power carbon dioxide lasers
Schawlow	Laser light
Giodmaine	The interaction of light with light
Herriott	Applications of laser light
Miller	Communication by laser
Lieth and Upatnieks	Photography by laser
Pennington	Advances in holography

76. J. M. Carroll, The Story of The Laser. (Dutton, New York, 1970).

77. L. Mandel and E. Wolf, editors, Selected Papers On Coherence and Fluctuations of Light. Two volumes. (Dover, New York, 1970).

78. D. Marcuse, Engineering Quantum Electrodynamics. (Harcourt-Brace, New York, 1970).

IV. WHAT ARE PHOTONS?

Corpuscular models of radiation have enjoyed a
strong advantage over wave theories because the dynamics of particles are simpler than those of waves in
a deformable medium. Newton favored corpuscles because he could not see how geometrical optics could
have validity as a limiting form of wave optics. It
required the later contributions of Young, Fresnel and
Hamilton in the early nineteenth century to overcome
this difficulty. More difficulties for the wave theory
come from the circumstance that the properties of the
aether medium are very strange, and not at all like
those of the mechanical models which were popular
until late into the nineteenth century. The wave theory
won a "decisive" victory in 1850 when the speed of
light was found to be lower in water than in air.
Nothing more was heard of light corpuscles until
Einstein's 1905 paper on the photoelectric effect.
From that time on, the light quantum (or photon) has
played a very strong role in the thinking of all physicists. This was almost inevitable before the development of the quantum theory of radiation in 1927.
Maxwell's wave theory was mathematically complicated,
and simply did not work for many problems with which
physicists were concerned. For example, light could
eject photoelectrons, and hence came in packets of
energy $h\nu$. To be sure, light could be diffracted by
gratings, and it was thought that in some vague way it
was guided by the Maxwell waves. In any case, most
physicists in the first quarter of the 20th century were
concerned with problems in which the corpuscular
attributes of radiation prevailed over its wave-like
qualities. This preference for a particle interpretation was much strengthened after the discovery of the
Compton effect. That fact is, of course, that there
was no dynamical theory of light quanta in any real
sense beyond the bald assertion that they were particles
with energy, momentum, polarization, and statistics,
which could be created and destroyed in elementary
interactions with matter. See A. H. Compton, Rev. Mod.
Phys. 1, 74 (1929) for an article giving an account of
the status of photon theory in the late twenties.

The development of the quantum theory of radiation
provides a much more satisfactory interpretation of
phenomena involving the interaction of radiation and
matter. The essential step for this theory was the
development of quantum mechanics in 1925-26 which gave a
fine interpretation of the stationary states of an

isolated atom. Heuristic recipes involving "transition currents" were devised to allow the calculation of the rates of the spontaneous quantum jumps between stationary states, i.e., calculation of the Einstein A coefficients.

By inserting suitable new terms into the wave equation to describe an atom perturbed by a classical external electromagnetic field, Dirac and others in 1926 dealt with processes of absorption and stimulated emission, i.e., calculated the Einstein B coefficients. The procedures just outlined constitute the semi-classical radiation theory. (See Schiff's Quantum Mechanics for a fuller discussion of this theory.)

It was shown by Wentzel in 1926 that this theory was also able to give an adequate account of the photoelectric effect caused by a classical electromagnetic field. The Einstein relation, $h\nu = \frac{1}{2}mv^2 + \phi$, giving the kinetic energy of the ejected electron, emerges from the calculation, although absolutely no reference is made to photons in the derivation. The photocurrent is proportional to the square of the electric field amplitude or to the intensity of the incident radiation (more accurately, the low frequency part of these quantities.)

If this form of quantum mechanics had been known in 1905, classical electromagnetic theory could have coped with the photoelectric effect and it would not have been necessary for Einstein to introduce his light quantum hypothesis. Of course, as mentioned above, the semi-classical radiation theory is only able to deal with spontaneous emission by an assumption foreign to the theory, but then it also becomes capable to derive the Klein-Nishina formula for the intensity of the Compton effect. The semi-classical radiation theory has no power to deal with the thermodynamics of black body radiation, or with quantum electrodynamic level shifts.

The quantum theory of radiation was introduced by Dirac in 1927. Both matter and the electromagnetic radiation field were treated as dynamical systems subject to quantization procedures. The electromagnetic field was expanded in normal modes of a cavity (plane waves for unbounded space). The hamiltonian for each normal mode of the radiation field is dynamically equivalent to that of a mechanical simple harmonic oscillator. The stationary states for a normal mode of frequency ν had energies of the form $(n + \frac{1}{2})h\nu$ with

n = 0, 1, 2, ... This could be interpreted in photon
language by saying that there were n photons in the cavity
mode. For a plane wave normal mode the momentum has
the eigenvalue nhν/c. Thus, some comfort is offered to
one brought up on Einstein's light quanta. However,
nothing like a more localizable particle is in evidence.
In fact, no satisfactory quantum theory of photons as
particles has ever been given. On the other hand,
the quantum theory of radiation seems to give amazingly
satisfactory accounts of a very wide range of radiative
phenomena, and there is no real need to have a photon
theory.

In order to describe the interaction of the matter
and radiation systems, Dirac introduced coupling terms
into the hamiltonian for the complete system. He guessed
these, using classical theory as a guide. The coupling
led to transitions between stationary states of the un-
perturbed atom-field system. Dirac calculated rates of
spontaneous emission, absorption and stimulated emission
and obtained just the values inferred by Einstein in
1916.

Dirac's discussion was followed by that of Heisenberg and Pauli, who treated the electromagnetic field in
a more general fashion. They turned up a difficulty
which arose from the gauge invariance of the theory.
This has been dealt with in many ways, most often by
those suggested by Fermi, and much later by Gupta and
Bleuler.

By 1930, Weisskopf and Wigner had given a satisfactory theory of the natural broadening of spectral
lines as related to the exponential decay of an excited
atomic state. However, Oppenheimer soon showed that this
theory led to infinite shifts of the atomic energy
levels. He made an attempt to obtain finite energy differences so that the spectral frequencies would not disagree so violently with experiment. Oppenheimer's
work was carried out before the negative energy states
of the Dirac relativistic wave equation were interpreted by positron theory. He would have had more success
with his subtraction attempts had the negative energy
states been filled with electrons in the vacuum state,
as was done by Weisskopf and Furry a few years later.

The quantum theory of radiation developed by Dirac
and by Heisenberg and Pauli involved very elaborate
formalism, and its notation was quite unwieldy. Fermi
gave some summer lectures at Ann Arbor in 1930 which

greatly simplified the theory, and showed how the quantum theory of radiation could cope with a number of historically important problems. Besides considering emission and absorption processes, and the Doppler effect, he demonstrated that radiation propagates with the speed of light, and he worked out the theory of a typical interference experiment. The lecture notes were published in the Reviews of Modern Physics in 1932, and are still well worth reading today. The 1936 book Quantum Theory of Radiation by Heitler further simplified the presentation, and treated many problems of importance in high energy physics, such as the photoeffect, Compton scattering, bremsradiation, pair production, etc.

Despite its many successes, the quantum theory of radiation continued to be plagued by divergence difficulties. The way to sidestep these was shown by Bethe in 1947, followed soon by relativistically covariant formulations of the renormalization theory by Schwinger, Feynman and Dyson. A Dover reprint volume edited by Schwinger in 1958 collects a number of these and other papers.

A good proportion of the physicists who carried out research on quantum electrodynamics received their university salaries for teaching classical electromagnetic theory, or even the theory of alternating current circuits. During the 1939-1945 war some of them did radar-directed research on problems of transmission and generation of microwaves. After the war, a series of 28 volumes was issued to describe the radar research at M.I.T. Nowhere in these volumes is any mention of microwave photons. Still less do books on power engineering refer to 60 cycle per second alternating current photons. Why did it happen that physicists tend to think only of photons in connection with the generation and interactions of light, from visible through x-rays and gamma rays, and only of classical constructs like electric and magnetic fields, voltages, currents and phases in low-frequency work?

It is well known that classical mechanics can be regarded as having validity in the limiting case where the de Broglie wavelength is very small compared to the other distances characterizing the problem. (The motion of the moon is never considered to be a quantum mechanical problem.) One way of showing this validity is to use the WKB approximation (Wentzel, Kramers and Brillouin). Another way is to form wave packet solutions

of the Schrödinger equation. These are time dependent
linear combinations or superpositions of stationary
state eigenfunctions of the system under consideration.
There is a general theorem that the centroid of any
wave packet follows the classical trajectory of a particle acted upon by the quantum mechanically averaged
force on the wave packet. In general, wave packets
change shape and spread with time, so that a classical
treatment of the problem is valid only in the limit
given above.

In the special case of a simple harmonic oscillator,
any wave packet is a periodic function of the time, with
period equal to the classical period. A gaussian wave
packet remains gaussian but, in general, its width
pulsates. For the special case of a packet of gaussian
shape with width equal to that of the ground state
simple harmonic oscillator wave function, the root mean
square width remains constant in time. The packet
oscillates back and forth with a certain amplitude and
phase just as a classical particle does in the parabolic potential field.

For such a coherent (i.e., sticking together)
wave packet the probability amplitudes c_n for the
stationary state wave function $u_n(x)$ exp-$in\nu t$ depend on
n in a simple way. The probability $|c_n|^2$ of finding an
energy $(n + \frac{1}{2})h\nu$ is given by a Poisson distribution in
n. If n_p is the most probable value, the amplitude of
the wave packet's oscillation is given by the energy
equation $\frac{1}{2} m (2\pi\nu)^2 A^2 = (n_p + \frac{1}{2})h\nu$, and the root mean
square spread in n values characterizing the packet is
given by $\sqrt{n_p}$. When $n_p \gg 1$, the wave packet has a
width much less than its amplitude of oscillation, and
so many different "photon" states are involved that it
is pointless to use the photon concept at all.

Since the electromagnetic field in a cavity is
dynamically equivalent to a denumerable infinity of
harmonic oscillators, one can immediately carry over to
the quantum theory of radiation the above results for
mechanical oscillators. The stationary states for a
single mode of the field can be called n photon states.
In addition, there are wave packet states which correspond very closely to the sinusoidally oscillating
normal mode amplitude of a cavity resonator.

This relation between the mechanical and electrical cases is fairly obvious, and any competent theoretical physicist who thought about the problem would have found the answer. However, for about 25 years the Schrödinger wave packet was not put to use in electrical problems. Israel R. Senitzky in the early fifties was an exception, and after 1963, Roy J. Glauber and others developed extensive theories of the coherent state. This theory is elegant and mathematically beautiful, but it has not, in fact, played a central role in the development of a quantum theory of the laser. It does, however, afford a general and often useful representation for any quantum radiation field.

In the history of atomic physics since 1914, great emphasis has been placed on stationary states. Although an excited state can undergo spontaneous radiative transitions to lower states, the process is typically a slow one, with a life time of the order of a million times a corresponding classical orbital time. In the quantum theory of radiation, the n-photon states also are energy eigenstates, so that the photon concept received some support from the importance of the atomic stationary states. However, in quantum electronics, the n-photon states do not have a long life time.

One of the basic postulates of quantum mechanics is the principle of superposition. If a system can be in a state with wave function Ψ_a, or in a state Ψ_b, then it can also be in a superposition or linear combination state $\Psi = a\,\Psi_a + b\,\Psi_b$, with the complex numbers a and b depending on the time, so that Ψ obeys the time dependent Schrödinger equation.

Such linear combination states play an essential role in microwave and nuclear spin resonance experiments, and in the theory of lasers. Consider two atomic stationary states u_a and u_b between which laser action occurs. Neither u_a or u_b has a quantum mechanically averaged electric dipole moment. However, the linear combination state $\Psi = a\,u_a + b\,u_b$ does have such a dipole moment, and this is a periodic function of the time with frequency $\nu = (E_a - E_b)/h$. A collection of such atoms in a laser cavity resonator can act like an antenna, which generates an electromagnetic field of frequency ν in the cavity. In turn, the field causes

the amplitudes a and b to change with time. The fields
thus make dipoles, and the dipoles make fields. The
semi-classical theories of the maser and laser make use
of self-consistency conditions to derive the amplitude and frequency of oscillation as a function of
various parameters of the system. There is no need to
speak of stimulated emission, or of a photon avalanche.
The semi-classical theory of the laser suffices to deal
adequately with nearly all questions of practical
importance on the operation of the device. A fully
quantum mechanical theory of the laser has, however,
also been developed so that any desired corrections to
the semi-classical results can be made.

Dirac has written in Chapter 1 of his book Quantum
Mechanics that "Each photon then interferes only with
itself. Interference between two different photons
never occurs." This statement has caused confusion. It
is known that two separated radio transmitters can produce interference effects. If one thinks that each
transmitter sends out its own photons, there is an
apparent contradiction to Dirac's statement. The difficulty disappears when one remembers that both transmitters are coupled to the modes of the same radiation
field. A photon is simply a particular energy eigen-
state of one such radiation mode. The problem can be
analyzed either with classical or quantum electrodynamics
as appropriate, but in the latter case a description
in terms of interfering photons is doomed to failure.

V.   WHAT IS STIMULATED EMISSION OF RADIATION?

In order to answer this question, we will recall
some of the various ways in which emission and absorption of light have been considered since the introduction of the electron theory of matter around 1900.
Consider the Thomson model of an atom, where an electron
is bound in a harmonic oscillator potential. According
to classical electromagnetic theory, such an electron
set into oscillation along the z axis emits radiation
and loses energy at a rate proportional to the fourth
power of the frequency ω of oscillation, and to the
square of the amplitude A of oscillation. The spectral
distribution of emitted radiation is a Lorentzian with
full width at half-height simply proportional to the
radiative decay constant of the oscillatory motion.
The emitted radiation has its electric vector plane
polarized in the z direction. The angular distribution
of the intensity varies as the square of the sine of the
angle θ between z and the direction of observation, and
the intensity falls off with the inverse square of the
distance from the atom. The electric field at any point
far from the atom is a damped sinusoidal function of
time, and its phase is simply related to the phase of
the electron's motion.

It is generally considered that the above radiation
is the classical counterpart of spontaneous emission in
quantum theory. In his text book, Schiff points out
that the classical and quantum intensity equations differ
only by a factor of two (and even this can easily be
made reasonable.) However, there are important differences in the meaning of the two similar results.
Classical radiation has a definite phase, correlated to
the phase of the electron's motion. For simplicity, we
discuss the case where the quantum atomic oscillator is
initially in the first excited state n = 1. The electron's
motion then has no phase associated with it, and neither
has the electromagnetic field. The quantum mechanical
averages of the electron's displacement, and of the
electromagnetic field at any place or time are zero.
We could, if we liked, solve a different quantum mechanical problem and get a well-defined phase of the electric
field. We would ask for the radiation emitted by a wave
packet, for which the electron's motion does have a
definite phase, and would then get a result corresponding more closely with the classical result. (It would
suffice to take for the wave packet a linear combination
of only the n = 0 and n = 1 states. This problem is not
usually discussed in textbooks, but it is easily treated.)

We could, alternately, take the classical solution, and apply it to an ensemble of simple harmonic oscillator atoms with random phases. Then the ensemble-averaged electric field would be zero, and the correspondence between the two theories would be improved. (Of course, in principle, there is no reason why we should take an ensemble of randomly phased atoms: it is legitimate to ask for the quantum mechanically averaged field produced by an ensemble of identically phased atoms.)

There are, however, still essential differences between the meanings of the two results. Classically, the radiation has an angular distribution, and it is taken to be obvious that two classical detectors at different points will both indicate the presence of radiation. In the quantum case, one has to specify more carefully the nature of the detectors of radiation. A photoelectric detector counts photons. In a sense, it makes the photons, and then counts them. A device something like a cathode ray tube measures electric field strengths. If two photoelectric detectors were placed at different angles, only one of them, and not two, could give a response, since the n = 1 atomic state can at most emit one photon.

A more striking difference is that the classical atom emits radiation continuously in time, although at an exponentially decreasing rate. In the quantum case, the <u>probability</u> that the atom remains excited decays exponentially in time. (It is essential to remember that in working with probabilities we must always be dealing with ensembles of similar systems, in similarly prepared initial states and with similar measurement procedures.) If for any member of the ensemble a photon is found at some time, the atom will then surely be in the n = 0 state, and no further radiation can be emitted.

According to the quantum theory of radiation, the wave function of the atom-radiation field system changes continuously in time. At t = 0, the atoms are in the n = 1 state, and the radiation field is in its vacuum state, i.e., all photon occupation numbers are zero. As time goes on, the wave function acquires increasing contamination terms representing a ground state atom and a radiation field with just one photon present in the totality of all modes of the free field. The discontinuous jumps of the old quantum theory are not seen. The only discontinuity is introduced by observation or measure-

ment. The theory does not really correspond to the idea
of von Schweidler, who in 1905 regarded emission (whether
of alpha particles, or of light quanta) as a completely
random or Markoff process. However, it is sometimes a
useful and convenient fiction to think of transition
rates in the language of the old quantum theory, if one
is careful not to take it too literally.

If a photon found by a detector has a momentum $\hbar\omega/c$,
the atom has received an equal and opposite momentum
recoil. If the atom is bound (lightly) in a solid, the
possibility exists that any part of the recoil kinetic
energy may excite sound waves in the solid, and as a
special case we have the Mössbauer effect, where all of
the atomic energy $\hbar\omega$ goes into the energy of the emitted
photon. We know from the theory of the Mössbauer intensity that the recoil momentum is transferred in a very
short time compared to the period of vibration of the atom
bound in the solid. We can even say: "We don't know
when the emission takes place, but when it does, it
happens suddenly." However, such tempting language is
not really appropriate for the problem, for the moment
of emission could only be determined by observation, and
there is no operational significance to a question about
when or how suddenly the emission takes place. If we
look for a photon, we have a certain probability of
finding one which is given correctly by the theory. If
we ask a sensible question of the quantum theory of
radiation, we get a sensible answer.

We turn now to absorption and emission of radiation
by an atom. For simplicity, I will formulate the discussion in mechanical terms. Let the atom be represented
by a simple pendulum. At present, we consider only
small amplitude oscillations so that the atom is a
simple harmonic oscillator. The frequency of free
oscillation is denoted by ω. We exert a sinusoidal
force of the same frequency (resonance) along the arc of
the swing of the pendulum. This force is supposed to
represent the effect of an electromagnetic field on the
atom. If as the pendulum swings the resonant force has
a direction which augments the motion, the pendulum
gains energy. We are doing positive work on the atom.
This corresponds to the process of absorption of radiation. We are not seeing in detail how the radiation is
absorbed, but infer that this occurs from energy conservation arguments.

If, on the other hand, the phase of the periodic
force is 180° in phase ahead (or behind) the pendulum's

motion, the amplitude of swing will steadily decrease
from its initial value, eventually reaching zero, and
then increasing indefinitely. For a while we will have
something like stimulated emission. The atom will be
transferring its energy (kinetic and potential) to the
electromagnetic field. This classical model for stimulated emission was known to Van Vleck around 1924, and
was discussed in detail by Heitler in his 1936 monograph
Quantum Theory of Radiation. However, it appears from
Heitler's calculations that this "stimulated emission"
process can go on only for a limited time; for longer
times the pendulum will act like an absorber. Furthermore, in practice we could hardly expect to find the
atoms in the proper phase relation with the field. As
Heitler shows, if the calculation is more realistically
carried out for an ensemble of randomly phased oscillators,
the induced emission disappears completely, and one has
only absorption. Atoms behaving like simple harmonic
oscillators do not make good masers. This would also
be true of quantum mechanical simple harmonic oscillators,
since all the energy steps are equal, and "there is
more room at the top".

The simplest problems in quantum mechanics involve
atomic systems with only two states, or at least only
two states that have to be considered. Such systems
readily give maser action if a suitable population inversion can be produced. It turns out that a simple
change in model suffices to give maser action for our
completely classical analogue. We merely let our
oscillators be simple pendulums, i.e., non-harmonic
oscillators. The problem of a forced simple pendulum
(known as Duffing's problem) is a basic one in non-
linear mechanics. Non-linear problems are much harder
than linear ones, and physicists have tended to avoid
them when possible. However, the real world is extremely
non-linear, and hence such problems are sure to be increasingly studied in the future. The development of
computers will be of great assistance in such work,
since analytical methods are not very successfully
applied to non-linear problems.

Consider an ensemble of randomly phased simple
pendulums, all swinging with a certain initial amplitude A_o and frequency $\omega(A_o)$. Because of its non-linearity
the simple pendulum is not isochronous, and its frequency
$\omega(A)$ decreases as the amplitude A of oscillation increases.
Let each of these pendulums be acted on for a time T by
a periodic force with a frequency ν slightly higher than

ω (A_o). Under these conditions, some pendulums will be
phased so as to gain energy from the field. In doing so,
their amplitude A increases, and the resonance condition
$\nu = \omega(A)$ is even less well satisfied than initially.
Before long, the pendulums will be out of phase with the
field, and the amplitude will reach a maximum and start
to decrease. Consider instead a pendulum whose initial
phase is 180° out of phase with the field. Such a pendulum
will give up energy to the field. As it does so, its
amplitude decreases, and its free frequency of vibration
increases. The resonance condition $\nu = \omega(A)$ is increasingly well satisfied, and the driving force is able to
produce an enhanced effect. The amplitude may even
fall below that corresponding to resonance, and some time
will elapse before a turn-around occurs.

Detailed calculations show that the average energy
of the ensemble pulsates, first rising, and then fal-
ling to a lower value than the starting energy. If the
interaction is broken off at a suitably chosen time T,
there will have been a net transfer from the atomic
system to the field. To get maser action, it is only
necessary to send a beam of such randomly phased classical atoms with amplitude A_o through a cavity resonator
tuned to frequency ν slightly above $\omega(A_o)$. If the
intensity of the beam exceeds a certain threshold
value, the energy transfer will more than make up for
cavity losses, and continuous maser oscillations will
occur in the cavity.

In the foregoing, we have concentrated on the effect
of the field on the atoms, and have only inferred from
energy conservation arguments that the field will grow in
energy at the expense of the atoms. By fairly complicated
calculations one can see just how the beam atoms
generate the electromagnetic field. (See Borenstein
and Lamb, 1972).

It is interesting to consider the refractive
properties of a medium made either of classical or
quantum atoms. Let us first consider the case of a
single classical simple harmonic oscillator atom,
initially without vibration. Let a plane polarized
electromagnetic wave of resonant frequency ω fall on
such an atom, and "turn on" the interaction at t = 0.
The field will set the atomic electron into motion and
the atom will generate electromagnetic waves. At large
distances from the atom there will be a clearly discernible
scattered wave, but very close to the atom the field

becomes singular. As we have seen in the previous discussion, this atom acts like an absorber because it
gains energy from the field. At large distances, the
field consists of the incident plane wave and outgoing
spherical wave. These may be represented schematically
by e^{ikz} and $f(\theta)e^{ikr}/r$ respectively, where $f(\theta)$ is
a certain function of angle. In most regions, it is not
hard to distinguish the incident and scattered waves.
However, very close to the forward direction, the two
waves are easily confused. A calculation of the fields
shows that there is a destructive interference between
the incident and scattered waves of importance just in
or near the forward direction. A calculation of the
Poynting energy flux density shows an incident energy
flow for the incident plane wave, a diminished energy
flow downstream from the scattering atom, corresponding
to the interference field, and a radial outward flow
corresponding to the scattered wave for $\theta > 0$. An overall energy balance would take into account the various
field energies, as well as the kinetic and potential
energies of the oscillator.

If the atomic oscillators are initially in motion,
it is possible to have either absorption or stimulated
emission, depending on the phase relationship. In the
latter case, one finds that the sign of the scattered
wave is reversed, so that the interference flux in the
forward direction reverses sign. This corresponds to an
amplification of the incident wave. We will not carry
out the discussion for the anharmonic oscillators, but
it should be obvious that they will generate a field
with the correct energy flow.

We now consider the case of a medium of such atoms,
distributed at random in a plane parallel slab. This
problem can be treated by adding up the waves sent out
by each atom. One finds that the medium is characterized
by a complex index of refraction, and the sign of the
imaginary part of the index of refraction changes when
one passes from the case of an absorber to that of an
induced emitter.

These problems can also be treated with the quantum
theory of radiation, and a quantum mechanical two level
atomic model. Of course, suitable quantum mechanical
averages of the field operators would replace the classical quantities. The results cast further light on the
nature of stimulated emission, but are fairly obvious
after an understanding of the classical theory.

VI. CONCLUDING REMARKS

One reason for presenting the Chronological Listing was just that it was so very long. It is amazing
how many people have contributed directly or indirectly
to the development of the laser and to its understanding.
One should also note that most of the contributions
consist of single items; only a lucky few have made
multiple contributions of importance.

One question often asked is "Why did it take so
long to invent the laser?" Garrett points out that Lord
Rayleigh might well have found it almost a century
ago. It could have happened that way, but there are
strong reasons why all sorts of developments in the
understanding of the nature of radiation and matter had
to occur first, in addition to advances in technology;
for instance of multi-layer interference mirrors, high
speed electronics, etc. The situation is no worse than
it was with the development of rockets for space flight
where the basic physics was contained in Newton's
Third Law of Motion. This paper indicates that a
similar technical lag occurred also in the case of the
laser.

ARTHUR L. SCHAWLOW

FROM MASER TO LASER

Arthur L. Schawlow

Department of Physics

Stanford University, Stanford, California

In some ways, lasers seem to be the realization
of one of mankind's oldest dreams of technological power.
Starting with the burning glass, which was known to the
ancient Greeks, it was natural to imagine an all-
destroying ray of overpoweringly intense light. Francis
Bacon, in his 1627 New Atlantis, imagined that the inhabitants of this utopia had "all multiplication of
light, which we carry to great distance, and make so
sharp, as to discern small points and lines." In War of
the Worlds, H.G. Wells' 1898 novel, Martians nearly
conquered the earth with a sword of light. In 1923, the
Russian novelist Alexei Tolstoi wrote The Hyperboloid of
Engineer Garin. Then, in the 1930's the Buck Rogers
comic strip often made use of a disintegrator gun.

Yet many old dreams, which have more or less come
true in this century, are realized only more or less.
Men dreamed of flying like birds and now they do fly,
but it is not at all like birds. Similarly, most lasers
deliver far less than the destructive death rays of
science fiction but their light has properties, such as
monochromaticity and coherence, which go far beyond the
old dreams.

Rays of any kind were far from the minds of Charles
H. Townes and myself when, in 1957, we began to think
seriously about the possibility of optical masers.
Rather, we were thinking of what was already a classic
problem in pure technology: to find something which

113

would act like a radio tube and generate shorter radio
waves. "Daedalus" has pointed out in <u>New Scientist</u>
(December 22, 1965) that there is a body of research
which seeks to find ways to do things for their own
sake. There may well be no immediate application in
sight, but such pure technology "like pure science,
often has to masquerade as the applied variety in order
to get funds." Some problems in pure technology may
appear as frivolous as "the development of a square
gramophone record played with such a perfect quadri-
lateral-linear motion that corner effects are imper-
ceptible." But others play a serious part in the
development of technology. Even though their applications are not immediately foreseeable, they do parallel
or extend lines of enquiry which have been fruitful in
the past.

     Throughout the twentieth century, scientists and
engineers have sought to extend radio techniques to
shorter wavelengths. As a boy in the 1930's I had read
in the Radio Amateur's Handbook that after World War I,
the amateurs "couldn't go up [in wavelength], but we
could go down. What about those wavelengths below 200
meters? The engineering world said they were worthless
--but then, they'd said that about 200 meters, too."
After preliminary tests and "some months of careful
preparation, two way amateur communication across the
Atlantic finally became an actuality when Schnell, 1MO,
and Reinarty, IXAM, worked for several hours with 8AB,
Deloy in France, all three stations using a wavelength
of about 110 meters." Still shorter waves, with lengths
ranging from 10 to 80 meters, were found to make possible
world-wide communications.

     In the 1930's, amateurs and others found ways to
use very high frequency waves whose lengths were shorter
than about ten meters. These waves did not travel
much more than line-of-sight-distances, but they were
found to be suitable for reliable, broad-band broadcasting such as for television or stereo music. With
inventions like klystrons and cavity magnetrons it became possible to explore the properties of waves of
centimeter lengths. These waves were not suitable for
broadcasting since they could be stopped by almost any
obstacle. However, their short wavelength made them
useful for high-definition radar and for relaying broad-
band communications.

     From all this, it seemed overwhelmingly probable

if some way could be found to generate shorter wavelengths, there would be uses for them. Some of the
uses would be obvious, like communications, but there
was a good chance that the unforeseen uses would be even
more exciting. There were, of course, ways of producing
shorter electromagnetic waves from many kinds of hot
bodies. Such sources, like the sun and electric lamps,
could be quite bright, but they lacked several of the
desirable properties of electronic oscillators. The
output was always a rather broad band of frequencies.
Since the excited atoms or molecules radiated spontaneously and independently of each other, their output did
not have spatial coherence. Moreover in the infrared,
and especially for the longer infrared wavelengths,
spontaneous emission was relatively slow so that the
power emitted was small.

One of the requirements for building an electronic
oscillator to generate such short electromagnetic waves
is the resonators to tune it. For microwaves, which
have lengths ranging from millimeters to centimeters,
tuning is usually achieved with some kind of cavity
resonator whose dimensions are comparable to the wavelength. When the desired wavelengths are a small
fraction of a millimeter, construction of cavity resonators becomes a very difficult task. But nature has
provided us with many kinds of atoms and molecules with
natural resonances throughout the infrared and optical
wavelength regions. Even when I was an undergraduate
student, in the late 1930's, it seemed to me that there
ought to be some way to use these in amplifiers or
generators of infrared waves. But I did not know
enough quantum physics to even begin trying to find a
way to do it. Very likely others had similar vague
ideas and, indeed, the formal similarity between atomic
absorptions and resonances of tuned circuits had long
been recognized.

The connection between radio waves and atoms was
again emphasized by the growth of radiofrequency and
microwave spectroscopy in the years after World War II.
I was then a graduate student at the University of
Toronto, having interrupted my studies for war work
teaching at the University and then microwave antenna
engineering in a radar factory. At the University of
Toronto, we did not have the facilities for the
glamorous fields like nuclear physics and radiofrequency
resonances. So, I was happy to work, under Professor
Malcolm F. Crawford, on hyperfine structure in the

spectra of atoms. With another graduate student,
Frederick M. Kelly, I constructed an atomic beam light
source to give spectral lines sharp enough so that
their hyperfine structure could be analyzed. Another
graduate student, William M. Gray, constructed a spectrograph and a Fabry-Perot interferometer to use with our
source. Thus I became highly familiar with this interferometer, consisting of two parallel, partially-
transmitting mirrors facing each other. This instrument
had been studied in undergraduate optics classes, but
even though most of the work with the interferometer
was done by the others in our group, I did learn more
about it during our research. When I began to think of
resonators for light waves a decade later it seemed
natural to start with the Fabry-Perot structure of two
mirrors facing each other.

In the postwar years, it seemed to me that the
most exciting physics research was at Columbia University.
I.I. Rabi was still active, and W. Lamb and P. Kusch had
recently made discoveries which were immediately recognized as important and later brought them Nobel
Prizes. I wrote to Rabi, and he suggested that I apply
for a postdoctoral fellowship to work with Associate
Professor C.H. Townes. This fellowship was provided by
the Carbide and Carbon Chemicals Corporation, a division
of Union Carbide, to support research on the applications
of microwave spectroscopy to organic chemistry. I had
neither knowledge of nor interest in organic chemistry,
but microwave spectroscopy was an attractive new field.
I must also confess that I had not heard of Charles
Townes, although I soon found that he had recently
published a number of discoveries. At any rate, I
applied for and was awarded the fellowship.

After coming to Columbia University, I learned that
although microwave spectroscopy can be used to determine
the structure of organic molecules and for analysis,
that was not the only reason for Carbide and Carbon
Chemicals' sponsorship. As early as August, 1945, Dr.
H.W. Schulz, a member of their research staff, had
written a memorandum to propose a new type of catalysis
"to employ electromagnetic radiation of a specific frequency to effect activation of reacting molecules by
induced resonance." In this memorandum he stated that
"The pertinent frequency range would cover the long and
short wave radio bands as well as the infrared, visible
and ultra-violet spectra. A literature search indicates
that this principle has previously been employed only in

the case of photocatalysis." As his study proceeded,
Dr. Schulz came to realize that resonance catalysis
would need the tunability and power of radio generators,
but at a shorter wavelength than was available from
existing oscillators. After various alternatives were
considered, it was decided to support long range research
aimed in this general direction at a major university.

At Columbia University, there was a Radiation
Laboratory group in the physics department, continuing
a program from the wartime days on magnetrons to
generate millimeter length radio waves. Also there was
Townes, who had recently come from Bell Telephone
Laboratories and was making pioneering studies of the
interaction between microwaves and molecules. The
laboratory was supported by a Joint Services contract
from the U.S. Army, Navy and Air Force, with the general
aim of exploring the microwave region of the spectrum
and extending it to shorter wavelengths. Dr. Harold
Zahl of the Army Signal Corps and Paul S. Johnson of the
Air Force Office of Scientific Research were among those
active in the sponsorship of this program. Captain
Johnson also organized a millimeter wave study committee
and asked Townes to be its chairman. As Townes has
recounted, it was on the morning of a meeting of this
committee that he conceived the idea of the maser.

Thus during my stay at Columbia there was wide-
spread recognition that it was interesting to find
better ways to generate wavelengths shorter than those
produced by existing electronic devices. But nobody
had a good idea of how to do it, and so Townes' group
concentrated on exploring the structures of molecules
and their interaction with microwave radiation. This
turned out to be the right decision, for a detailed
understanding of the ammonia molecules was just what
Townes needed to invent the maser in the Spring of 1951.
In this ammonia maser, a beam of ammonia molecules
would pass through a suitable electric field which would
accept those in excited states and reject the unexcited,
absorbing molecules. The excited molecules could be
stimulated to emit microwave radiation inside a cavity
resonator, a metal box having dimensions comparable
with the wavelength.

Townes told me about his idea in May or June of
1951, and it seemed promising. I would have liked to
work on it, but my time at Columbia University was coming
to an end and I had accepted a job in solid state physics

at Bell Telephone Laboratories.

My work took me quite far away from problems of
generating electromagnetic radiation. I kept in touch
with Townes, because we were writing a book on Microwave Spectroscopy and I spent nearly every Saturday at
Columbia University. Thus I heard from time to time
about the problems and progress of the work on the
maser and was delighted when it first operated in 1954.
At about that time, interest in masers began to pick up
at Bell Telephone Laboratories and two years later, G.
Feher, H.E.D. Scovil and H. Seidel built the first
three-level solid state microwave maser following a
proposal by Nicolaas Bloembergen of Harvard University.
While the original ammonia maser had been primarily
useful as a frequency standard or as a sensitive
detector for studies of the ammonia molecules, the
solid state maser was something that could actually be
used for communications and radar. It had a broader
band width and could be tuned by changing the strength
of a magnetic field. Not long afterwards, C. Kikuchi
of the University of Michigan showed that ruby was a
good material for such masers. Joseph Geusic, who came
to Bell Labs from Ohio State University about that time,
where he had done his thesis with J.G. Daunt and had
for the first time measured the mivrowave resonances in
ruby, was one of those who became active in designing
and perfecting ruby masers.

I did not participate in any of this except as a
spectator, being busy with research on superconductivity
and, for a time, nuclear quadrupole resonance. I also
taught twice a three month course in solid state physics
for the engineers in the program which Bell Laboratories
had established for new engineers coming from college
with Bachelor's or Master's degrees.

When parametric amplifiers were rediscovered by
Harry Suhl, we thought perhaps this might somehow be a
clue to producing shorter wavelengths and I spent a
little time learning about them. I even built an audio-
frequency parametric amplifier too. It may well have
been the first one at Bell Labs since the work of
Peterson some years earlier, which had by then been
nearly forgotten and was not known to me.

By 1957, I was coming to think that the time was
right for a serious investigation as to whether one
could build some kind of an infrared maser. Naturally,

I was thinking primarily about the wavelength region
just a little shorter than could be obtained by radio
tubes.   Townes had hoped initially that his ammonia
maser would oscillate at a wavelength of a half millimeter, but in the final device the output was at one
and a quarter centimeters wavelength, which is well
within the region spanned by existing microwave tubes.
I remember attending a conference on low temperature
physics at the University of Wisconsin in August 1957
and chatting there with Michael Tinkham, who was then
at the University of California and had been doing some
far-infrared spectroscopy.   This kind of spectroscopy
was very difficult, because the existing light sources
were extremely weak and so I suggested that it really
ought to be possible to·build some kind of a maser to
produce a stronger source.   Tinkham mentioned that iron
in crystals had energy levels in a right wavelength
region, but neither of us did anything more about it at
that time.

        A few weeks later, about October of 1957, Charles
Townes visited Bell Labs and we had lunch.   Townes had
been consulting with the Laboratories for about a year,
but his contacts were with the maser people and I had
not had any serious discussions with him.   He told me
then that he was interested in trying to see whether an
infrared or optical maser could be constructed, and he
thought it might be possible to jump over the far infrared region and go to the near infrared or perhaps even
visible portion of the spectrum.   He had made some notes
and said that he would give me a copy.   We agreed that
it might be worthwhile for us to collaborate on this
study and so we began.

        We both realized that the three-level and four-
level pumping schemes, used in microwave masers, could
be used with incoherent light as a pump if we could get
enough power from the incoherent light.   Indeed, Townes
had envisioned optical pumping of masers as early as
1954, and had mentioned the method in his basic maser
patent.   Just as in the microwave maser the ammonia
molecules are excited independently and enter the
resonator quite individually, so we could excite individual atoms or molecules in any kind of a maser at
random.   The synchronization would be achieved by a
wave stored in a resonator.

        However, excited atoms lose their stored energy
even if they are not stimulated.   In solids at

radiofrequencies the energy is lost by transfer to the
crystal vibrations where it becomes heat, but in the
visible region spontaneous radiation may be more important. It was not at all obvious whether one could get
enough excited atoms despite spontaneous emission, and
the only way to answer this question seemed to be to
study the properties of some fairly simple substances
which might be calculable. Although solids and liquids
are known to emit strongly, gases are simpler and better
understood and the simplest gases are those consisting
of individual atoms. Townes thought he saw a suitable
system in thallium vapor and had described it in the
notes he gave me.

The thallium atoms would be excited from the
ground state (6p) to a higher one (either 6d or 8s) by
ultraviolet light from a thallium lamp. Such lamps
were in use in Kusch's laboratory at Columbia University
for experiments on optical excitation of thallium atoms
in an atomic beam resonance experiment. Townes had discussed with Gordon Gould, a student of Kusch's who was
working on the atomic beam experiment, the properties
of thallium lamps to find out how much power could be
expected from them. Atoms excited to the 6d or 8s
level would, according to Townes' scheme, rapidly
radiate part of their stored energy and drop to the 7p
level which would be the upper level for maser action.
From there they could be stimulated to make transitions
to a 7s level which would normally be empty.

After looking at this, I saw a flaw in it, in that
the rate of spontaneous transition was greater out of
the 7s to the 6p than from the 6d or 8s into the 7p.
This means that the 7p state, which was to hold atoms
to be stimulated, would empty faster than it filled.
Laser action might not be impossible under those circumstances, but it would be difficult, and it pointed out
a general problem with this sort of a cascade operation
in atoms. It is rather usual, although there are exceptions, that the various excited states have progres-
sively longer lifetimes as you go up except for the
ground state whose lifetime is essentially infinite.

However, if Townes' thallium scheme was not
immediately workable it did make an important point.
It would be easier to do a theoretical analysis for
transitions of a maser to emit radiation in or near the
visible than it would be for the submillimeter region,
where so little was known experimentally. It might even

be that it would be actually easier to build one in the near-visible region because the spacings between energy levels in that region are large enough so that thermal excitations do not quench excited atoms as quickly as they do for levels with the smaller spacings corresponding to the far infrared. So, we searched for suitable energy levels and transitions in some atoms which might be excited to emit radiation in this portion of the spectrum.

In this quest, we had a good deal of information to guide us, although not all the questions we wanted had been asked. The energy levels of many atoms were tabulated in the volumes prepared by Charlotte Moore of the National Bureau of Standards. Some transition probabilities were given in the Landolt-Börnstein Tables, in a Table edited by L. Biermann. These would give us a start and would give references to more complete information in original papers.

How many excited atoms would we need? Townes had the maser equation which he modified by letting spontaneous emission loss replace the other kinds of losses which had been dominant for microwave masers. The equations are given in the paper which we published in 1958, but essentially what he did was to imagine light waves traveling in a box which could be thought of, for the derivation, as a rectangular box with reflecting walls. Light would be lost only at the walls, and, knowing that light waves travel at the velocity of light, one could easily calculate the average time between wall reflections. Then from that you could get the ratio of energy lost to energy stored for an electromagnetic wave in the box, provided you know the reflection loss each time the wave reaches a wall. The rate of stimulated emission of energy from the excited atoms depends on the intensity of the stored wave, as do the losses. One needs then to calculate how many excited atoms are needed to overcome the losses. Now the excited atoms will radiate in a short time which might range anywhere from billionths of a second to perhaps thousands of a second and must be replaced on the average once each lifetime. Thus we can calculate the number of excited atoms needed per second to just make up the losses in this resonator. If we had more than that we can increase the losses by opening the hole or making the walls partly reflecting so that we can take out some of the energy generated.

In the microwave regions, the strength of the

interaction of the molecules with the stored electromagnetic wave is usually measured by the dipole moment
of the molecule. One can give an effective dipole
moment for an optically excited atom, but it is more
usual to use the quantity known as the oscillator
strength, f, which is related to the dipole moment.
This is the quantity most commonly tabulated in places
such as the Landolt-Börnstein Tables and more extensive
compilations which have appeared since then. The
oscillator strength, f, indicates the effective number
of electrons available for the particular atomic resonances and may range from one down to a very small
fraction of one, or in a few exceptional cases it can be
greater than one but not commonly. It can be measured,
for example, if you know how many atoms there are in the
ground state by determining the strength of the absorption of light within the band which the atom can
absorb. Measurement for excited state is more difficult because it is not easy to know just how many atoms
there are in each of the excited states but some measurements have been made.

Probability of stimulation by a given wave is
proportional to oscillator strength, and so also is the
gain for a given number of excited atoms. Thus to get
a large gain without exciting very many atoms, we would
wish to have a large oscillator strength. However, since
the oscillator strength measures the interaction between the atom and an electromagnetic wave, the rate of
spontaneous emission is also proportional to the oscillator strength. That is, the greater the oscillator
strength the shorter the lifetime of the excited state
and the faster we have to replace the excited atoms. It
turned out, then, that it did not really matter what
the oscillator strength was for the particular transition. If it was high, we would need only a few atoms
but we would have to replace them frequently. If it
was low, we would need many atoms, but would not have
to replace them as often. Thus the oscillator strength
would not matter at all, if atoms lost their excitation
only by emitting the desired radiation. But if there
are competing processes, it is helpful if the desired
one has a large oscillator strength.

Another factor, important for the gain of a
particular atomic resonance, is the width of the spectral line. The probability of stimulated emission,
and hence the gain from a given number of excited atoms
is inversely proportional to the width of the spectral

line. Fortunately, for a gas at low pressure, the
linewidth is known to be given by the Doppler effect
from the thermal motions of the atoms. This is easily
calculable. In solids and liquids the linewidths are
much more variable. When, later, we began to think
seriously about these materials, we had to make our own
measurements of linewidths.

We concentrated our study on the simplest atoms,
the alkali metals. While the hydrogen atom's spectrum
is perhaps even simpler and more theoretically calculable, hydrogen exists in the form of molecules which
have to be disassociated and the efficiency of the
dissociation would introduce additional uncertainty. The
alkalis have only one electron outside a closed shell
and so can be thought of as nearly one-electron atoms.
Their energy levels are well known and the metals are
not hard to vaporize. Moreover, alkali vapor lamps are
commercially available by a number of companies and indeed sodium vapor has been widely used for street light-
ing. I chose to look most carefully at potassium for a
rather trivial reason. Both the first and second members
of the principal series of potassium vapor lie in the
visible region. That is, one could pump potassium atoms
from the ground 4s state up to the 6p with visible
4047Å light from a potassium vapor lamp and then monitor
the progress of these atoms back to the ground state by
looking at the red line emitted when atoms drop from the
5p to the 4s state. In the other alkalis, one or the
other of these transitions lies in the infrared or
ultraviolet. These are obviously not very important
considerations, but I had essentially no optical
equipment at all at the time and was thinking whether
one could begin experiments easily and cheaply. Moreover, it did seem that any conclusions from one atom
would be pretty much applicable to the others.

I bought some commercial Osram alkali vapor lamps
and one of my colleagues, Robert J. Collins, measured
the power output of some of these lamps for me. Collins
had done his thesis research in spectroscopy and was by
that time a Bell Laboratories physicist working on some
infrared spectroscopic studies. He found that each of
the lamps generated to 0.08 milliwatts in the 4047Å line.
Of course this was only one small lamp and was not
designed for maximum power. One could imagine buying
large arrays of such lamps or, if necessary, building
them. But the 0.08 milliwatts, if we could use all of
it, would be sufficient to excite quite a large number

of atoms in a potassium vapor cell. So as our calculations progressed, with the aid of tables of measured
oscillator strengths published years before, it began to
look that you could indeed get enough excited atoms to
obtain measurable amplification in the excited state.

        During this time we had not been paying too much
attention to the resonator which we would need to
complete the maser oscillator. I had in mind from the
beginning something like the Fabry-Perot interferometer
I had used in my thesis studies. I realized, without
ever having looked very carefully at the theory of this
interferometer, that it was a sort of resonator in that
it would transmit some wavelengths and reject others.
Such an interferometer might typically have had mirrors
with diameters of perhaps 7 cm and spacing of perhaps
that much or less. Somehow it must have been implicit
in our thinking that the absence of the side walls did
not really matter too much. However, as we began to
feel satisfied that it was possible to get sufficient
excitation our attention turned more toward the properties
of the resonator. The number of modes of oscillation
of such a resonator, having dimensions tens of thousands
of times larger than the wavelength, was enormous even
in the limited range of frequencies which the atoms
could amplify.

        All physicists learn how to calculate the number of
modes of waves in a large volume somewhere around the
end of undergraduate or the beginning of graduate studies.
This kind of calculation is important, for example, for
estimating the spontaneous emission lifetime of excited
atoms and for the derivation of the well known law for
the intensity for emission from a heated black body.
The same kind of counting up of modes is used in the
Debye theory of specific heat, where the thermal motions
of the atoms in a solid are considered to be entirely
equivalent to a superposition of all possible random
thermal waves of wavelengths from very long ones to those
whose wavelength is just twice the spacing between the
atoms. I have been through this as an undergraduate,
but my memory had been particularly refreshed when I
taught the Debye theory of specific heat to the engineers at Bell Telephone Laboratories. However, at
first I simply looked at the number of modes and then
began to think what the output of the optical maser
might be like if we had one.

        Martin Peter, another colleague at Bell Laboratories,

was particularly insistent that we should find some way
to reduce this enormous number of possible modes. Otherwise, he felt the optical maser, if it did oscillate,
would jump rapidly from one mode to the other and not
produce any very recognizable kind of oscillations.
The coherence of the radiation would be continually
interrupted by jumps from one mode to a different one.
Townes had recognized the importance of this multimode
problem, and it had kept him for a long time from
proceeding with short-wave masers. When we began our
work together, he believed both that it was important
to damp out other modes and assure that there was good
mode control, but that even though he could see no
system which would do this completely, one should go
ahead in any case, thinning out the modes as much as
ideas would permit. He expected that the oscillator
would oscillate momentarily on a single or a limited set
of modes because of nonlinearities, but that it would
also jump fairly rapidly between different modes. He
believed that one could easily determine that the system
had gone unstable and was working, and that the properties
even with a complex set of modes would be recognizable
and interesting. Very possibly, if Townes and Peter had
discussed the question directly, they would have reached
some sort of agreement. None of us doubted that some
good method of mode selection was highly desirable.

I began to think of these modes in terms of the
waves of the Debye picture, that is waves traveling in
different directions inside the resonator and having
different wavelengths. The range of wavelengths was
limited by the bandwidth of the amplifying atoms. Now,
to reduce the number of directions that would be acceptable in the instrument was not so easy. I thought for
awhile that perhaps there might be certain directions
in which the light could come out of the box, as there
is in a Fabry-Perot resonator, and that the output might
be an array of beams like the Laue spots of the x-ray
diffraction camera. I thought at one time of replacing
the walls of the box by diffraction gratings ruled so
that they would only reflect light well for a particular
angle of incidence and that only waves coming in this
right direction would be properly reflected.

Having advanced this far, around the beginning of
February 1958 I wrote down my ideas about optical masers
in my notebook. Of course many very wise scientists
will tell you that any scientist worth his salt carefully records all observations, calculations and concepts

in his laboratory notebook. However, I fear that I do
not qualify, because in seven years at Bell Telephone
I had not yet filled one notebook. Indeed my experience had been that the only valuable calculations
and data were those that I took on scraps of paper.
Whenever I thought I had things in sufficiently good
order to record them in the notebook it turned out that
I had overlooked something and that that particular work
was essentially worthless. However, I did write down a
number of pages of thoughts about optical masers. They
included some calculations on potassium and the re-
ordering of the equations and some of the ideas about
possible structures, even though I was not at all con-
fident that the problem of mode selection was solved.
Even though I had never tried to patent anything, I
asked Solomon L. Miller to read and witness these notes,
on January 29, 1958. Miller had been one of Townes'
graduate students when I was at Columbia University and
had a laboratory near mine at Bell Telephone Laboratories. He was certainly well able to understand the
discussion in my notes and so indicated when he signed
them. I was a bit startled, just a few days later, to
learn that Miller had left Bell Labs to go to IBM.
Perhaps that had something to do with my never writing
any more ideas in my notebook.

But indeed I did get a good idea very soon after
writing these notes. I realized that if we took literally the Debye picture in which the various modes were
waves having different lengths going in different directions, it could suggest a way to select one, or at
most a few of these modes. If a wave started from somewhere near one wall of the resonator, it would reach a
different place on the far wall, depending on its direction. If, therefore, most of the far wall were
eliminated so that only a small patch remained, the wave
would only be reflected if it were going in the right
direction to reach that small patch of wall.

Thus we could reduce the large box to just two
small mirrors facing each other at the ends of a long
column of excited atoms. This arrangement would serve
as a good resonator for waves which travel nearly
straight along the axis joining the mirrors. A wave
with any other direction would soon move sideways enough
to miss the small end mirror, and thus would be lost.

It was clear to me then that this resonator could
not hold any wave unless its direction of propagation

was inclined to the axis by less than the angle subtended by one mirror at the position of the other.
Townes pointed out that this structure would be considerably more selective than that. Waves were expected
to bounce many times back and forth through the amplifying medium. Only a wave traveling quite exactly along
the axis would remain in the amplifying medium long
enough to attain a high intensity by stimulating emission.

     This simple reasoning convinced us that we had
found a structure which would really strongly favor the
growth of a few selected modes. It was also apparent
that the output through one of the partially-reflecting
mirrors would be a highly directed beam, more or less
approximating a plane wave. We were also satisfied that
we knew of at least one substance in which we would be
able to excite enough atoms for optical maser action.
However, it would take an uncertain time to build one,
and unexpected experimental problems might well be encountered. We were aware that, during the three years
which it took to construct the ammonia maser, some of
the ideas had been discovered and published by others.
There were many more workers in the field by 1958. Although we were not aware of any direct competition and
did not particularly try to hurry, it seemed best to
publish our conclusions without waiting for experimental
verification. During the spring months we worked mostly on writing the manuscript.

     Before submitting the paper for publication, we
were required to circulate the manuscript to our
colleagues at Bell Telephone Laboratories for technical
comments and to the patent department to see if it involved a patentable invention. Several people, particularly some of those most expert in microwave waveguide
theory, were skeptical of the reality of our modes and
the proposed method of mode selection. They wanted to
see a more complete calculation with rather precisely
defined boundary conditions, which was done only later,
in 1960, by A.G. Fox and T. Li. We did, however, add
some paragraphs to our paper, in the hope of making the
mode frequency and selection argument more complete and
clear. There was some worry that there might be some
modes of the resonator with longitudinal field components,
as are found in microwave resonators. However, in a
resonator like ours, the wave travels many thousands of
wavelengths from one mirror to the other, and must behave much like a wave in free space, and so must be
largely a transverse wave.

The patent department was, at first, quite un-
interested in the idea. I suppose that it appeared to
be remote from the needs of the telephone industry and
perhaps they did not believe it would work or that if
it did it would be very useful. However, largely at
Townes' insistence they did prepare and file an application for a patent. It was issued rather speedily, in
March, 1960. Our paper was submitted for publication,
in August of 1958, and was published in the Physical
Review in the December 15 issue of that year.

The paper did arouse a considerable amount of
interest and a number of laboratories began searching
for possible materials and methods for optical masers.
Townes, in his own group at Columbia, began efforts to
construct a potassium optical maser, working particularly through two graduate students Herman Z. Cummins and
later Isaac Abella. They were joined for a time by
Dr. Oliver S. Heavens who is now Professor of Physics
at York University in York, England and who was even
then a world renowned expert on highly reflecting
mirrors.

We were of course aware of other possible materials
for optical masers. One of these was cesium vapor.
Cesium had the additional advantage that it could be
pumped by a strong spectral line from a helium lamp,
which happened to coincide with one of the cesium atom's
absorption wavelengths. This coincidence had been
noted in 1930 by C. Boeckner and mentioned in A.C.G.
Mitchell and M.W. Zemansky's book Resonance Radiation
and Excited Atoms (Cambridge University Press, 1934).
We noted in our paper that a cesium infrared maser could
thus be pumped by a helium lamp. This kind of a laser
was constructed and successfully operated by S. Jacobs,
G. Gould, and P. Rabinowitz in 1961. Thus by 1958 we
knew a number of gases suitable for optical maser action,
although we could not be sure which would be easiest.

Being at Bell Laboratories, I had been pretty
thoroughly indoctrinated to believe that anything that
you can do in a gas can be done in a solid and can be
done better in a solid. I therefore began to explore
the possibility of solid optical maser materials.
Albert Clogston, who was my immediate boss at Bell
Laboratories, had encouraged my interest in optical
masers and now encouraged me to, if I wished, drop superconductivity entirely and begin studies of possible
optical maser materials. On the other hand, nobody ever

suggested that we try and organize a group to build an optical maser. Anything I did I would have to do myself. There was a nearly invariable custom in the physical research department that each man was to be an individual scientist, and not an assistant to anyone else.

About the optical properties of solids, indeed my ignorance was quite total. However, even before our paper was published I began to learn a little bit. One thing that impressed me was that some materials such as ruby had broad absorption bands and narrow emission lines. Thus we were able to say in our 1958 paper that "The problem of populating the upper state does not have as obvious a solution in the solid case as in the gas. Lamps do not exist which give just the right radiation for pumping. However, there may be even more elegant solutions. Thus it may be feasible to pump to a state above one which is metastable. Atoms will then decay to the metastable state (possibly by nonradiative processes involving the crystal lattice) and accumulate until there are enough for maser action. This kind of accumulation is most likely to occur when there is a substantial empty gap below the excited level."

When writing that, ruby seemed like a tantalizing possibility because it did glow so brightly almost no matter how you excited it. Several people about that time had become interested in the optical emission from ruby including Saturo Sugano and Y. Tanabe and their associates in Japan, Irwin Wieder at the Westinghouse Research Laboratories, and Stanley Geschwind at Bell Laboratories. But ruby seemed also to present a very serious difficulty. The transition for the only strong fluorescence lines were absorbed by unexcited atoms in the same material. That is, they were resonance lines and so the atoms could both absorb and emit the same wavelength. Thus one would start out with the disadvantage that initially all the atoms would be absorbing, so that half of them would have to be excited before any amplification at all could be obtained. Without doing any calculations this really seemed like too much of an obstacle to overcome, although there might be some way because of the broad absorption bands for pumping. Several people made measurements on the fluorescence efficiency and strangely enough they all gave estimates ranging from 1 to 10 percent. I now think they were all trying to be conservative, but if we had done a calculation and believed their figures we

would have confirmed our prejudice that it was not
possible to obtain optical maser action in the resonance
lines of ruby. But I thought that perhaps there might
yet be some way, such as by splitting the energy levels
in a large magnetic field so that one sublevel at least
would be empty at low temperatures.

How was it that so many people at nearly the same
time began to study the optical properties of ruby?
Well, the one reason was that there had been an advance
in the understanding of crystals related to paramagnetic
resonance studies which could carry over into the optical
spectra. The dominant influence in my case and perhaps
in some others, was that ruby was being used in microwave masers. It was possible to visit Joe Geusic or
others at Bell Labs and find a drawer full of rubies
from which you could easily borrow samples. So much was
my practice to borrow samples that I remember a year or
so later that George Devlin marked on one of the spectra
"ruby ALS bought!," because up to then I had been mostly borrowing.

One advantage of all this interest in ruby was that
crystals had been ordered and were available and with
various concentrations of chromium in aluminum oxide.
Although I did not think that ruby was going to be any
use for an optical maser material, it was interesting
because it seemed to have a fairly simple spectrum,
which according to the theory should have had just two
emission lines at low temperature along with some bands.
Indeed at the very lowest temperatures it should have
only one emission line. In actual fact, as had been
shown many years ago by a number of experiments and most
thoroughly by Otto Deutschbein in 1932 and by S.F.
Jacobs and G.H. Dieke in 1956, there were very many
lines in the spectrum of ruby. I thought naively that
these extra lines might be due to the interaction of the
chromium ion with the crystal lattice vibrations and
that if we studied them it would give us some information
about the crystals. It also seemed interesting to look
at the chromium spectrum not only in aluminum oxide but
in the related gallium oxide which Joe Remeika could
grow as small crystals with various concentrations.

I had at that time a Gaertner wavelength spectrometer, a very simple student-type instrument which I
had bought when I first went to Bell Labs thinking that
it could be used for measuring thickness of thin metal
films. Darwin Wood was in charge of spectrochemical

analysis and he could make available some time on his
spectrographs and indeed collaborated with us on some
of our early studies.

I have mentioned George Devlin, who was my technician during most of my years at Bell Labs. I had
hired him even though he had very little formal training and really a rather poor high school education. He
had, however, been a champion model airplane builder
and it was evident that he had an attractive personality
and a quick mind. It turned out to be a very rewarding
association. For Devlin, although he was almost completely nonmathematical, had a real physical insight as well
as skill in designing, building and operating equipment.
Perhaps one of his most valuable characteristics, was
that he did not have any great preconceptions as to what
the data should be. Several times, he pointed out to me
small effects which I would have dismissed as noise but
which he insisted were real and turned out to be interesting. One of these was that when we were looking at
various samples of gallium oxide crystals with various
concentrations of chromium, he noticed that the satellite lines, that is the extra lines to the red of the
strong R-line, were different in different samples. That
was all I needed and I immediately jumped to the conclusion that these lines were not due to the crystal
vibrations but rather to pairs of chromium ions. The
probability of a given ion having a near neighbor in a
particular crystal ion position was going to be proportional to the concentration of the ions in the crystal.
At low concentrations such close neighbors would be
very unusual and at high concentrations they would be
common. This of course would also be the explanation
for the extra lines in ruby. Darwin Wood and I investigated this point more quantitatively, collaborating
with Albert Clogston on the theoretical aspects, and we
published a note on the pair spectra in the summer of
1959.

However, one of the most interesting features of
this pair spectrum was that it could produce a large
splitting in the ground state of the chromium ions
which the crystal field alone could not do. Thus instead of having a single or a very narrowly split ground
state the splitting could be large enough that at low
temperatures some of the higher levels of the ground
state would be empty. This was what we thought we
needed for an optical maser. We really felt that if we
were to get optical maser action we had to give ourselves

every advantage and that is why we had not seriously considered the theoretical possibility that we could empty out the ground state by pumping. I presented this dark red ruby as a possible laser material, which would oscillate in the satellite line at 7009 or 7041Å or perhaps both, at the First International Quantum Electronics Conference in September of 1959. In emphasizing the advantages of the broad pumping bands in ruby and the four-level system made possible by the exchange coupling in dark ruby, I said briefly that the R lines were not suitable for laser action. Of course, it turned out later that both sets of lines can be made to lase.

The proceedings of that conference, including my talk, were published very quickly and issued in February of 1960, by the Columbia University Press. This is one of the very few occasions when I remember exactly when I wrote a paper. I had promised a manuscript by the time of the meeting, but as so often happens other things had prevented it. Therefore, I stayed at home for the first two days of the three-day meeting, wrote the paper and then came and delivered my talk on the third day.

In that particular paper, I also described rather concretely the structure of an optical maser in the following words -- "The structure of a solid-state maser could be especially simple. In essence, it would be just a rod with one end totally reflecting and the other end nearly so. The sides would be left clear to admit pumping radiation."

Well, if we knew the material and the structure, why not do it? Now, when we know that construction of a laser can be so easy, it is hard to give a convincing answer to that question. But when we did not know how to assess the difficulties and since it had never been done, we believed that they might be formidable. For example, no solid, and especially not dark ruby, is free from variation of refractive index caused by strain. Thus if a plane wave started out at one mirror and was amplified as it passed down the rod, it would be distorted beyond recognition before it could reach the other end and would not return nicely to the first mirror. Nevertheless, I did manage to find from Bill Mims, who was working on microwave masers, a rod of dark ruby. And as early as December 1958 I had the ends polished flat and parallel. I still have the order to

have this done at Laboratory Optical Company. However,
I did not acquire flashlamps and merely tried this half
heartedly with a General Radio Strobotac which I had
bought for measuring fluorescence lifetimes. This was
not enough power and nothing happened. One other reason
I did not push more agressively on it was that I was not
sure just how cold the crystal would have to be to empty
the lower state of the optical transition. And besides,
there were too many interesting things to do in studying
the spectrum.

But others were active. At Bell Labs, Ali Javan
had conceived the idea of a helium-neon maser making use
of transfer of energy from metastable helium atoms
produced in a gas discharge to particular excited levels
of neon atoms. Actually, I had heard the idea of using
a gas discharge even before we wrote our first paper.
Willard S. Boyle at Bell Labs had mentioned this possibility but he did not explore it seriously. He in fact
described it in the context that processes in a semiconductor were similar to those in a gas discharge, and
that it would be interesting to try and find a semiconductor analog of a gas discharge and that might produce population inversion which would permit optical
maser action. Indeed Boyle has a patent on semiconductor lasers. We did not mention this idea in our
paper although both the gas discharge and semiconductor
possibilities seemed real, because these were Boyle's
ideas and not ours and it was up to him to publish them,
but he did not do it.

Ali Javan, however, did work out in some detail the
properties of a particular system and he published a
theoretical analysis in 1959. John Sanders from Oxford
who visited Bell Labs in 1959 for eight months or so had
another proposal using pure helium. People objected to
Sanders' scheme in that the lower level would not have
a short lifetime because of trapping of the radiation
which was supposed to empty it, but there were objections
in nearly everything anyone proposed and Sanders did
spend some time trying it out. However, he had to return to England before any real conclusions were reached.

Javan, whom I had known as a student with Charles
Townes at Columbia and later as a postdoctoral worker,
was and is an extremely ingenious and able scientist.
His enthusiasm attracted others, most particularly
Donald R. Herriott and he was able to arrange for the
Laboratory to hire William R. Bennett, Jr. They made

detailed studies of the processes in the gas discharges
and Herriott developed optical components of great
precision and quality for a helium-neon gas discharge
optical maser. This all took considerable time, and indeed the research management at Bell Labs became concerned whether this was all a waste of a rather considerable amount of money or whether there was indeed
some hope in it. I remember people went around and
asked various opinions, but since these opinions were
quite uniformly optimistic, the research was continued.
But Javan and everyone else believed that the conditions
for maser action in gases might have to be quite special
as there are many processes tending to restore thermal
equilibrium. I don't think anyone realized, as we now
know to be true, that nearly any gas will lase if excitation is violent enough. Indeed no one even considered looking at pulsed gas discharges until considerably later. Perhaps that was because of the pre-
occupation with communications around Bell Labs, for
which a continuous laser seemed the only really interesting one. Indeed that was another obstacle which detracted from my pursuing pulsed laser operation in ruby.

     Around the same time, C.G.B. Garrett and Wolfgang
Kaiser, both of whom had worked previously on semiconductors, became interested in trying to develop a
solid state optical maser. They were, very reasonably,
attracted to the rare earth compounds and transparent
crystals which, as we had noted in our paper, give strong,
sharp fluorescent lines. With these materials, four
level systems with any empty lower state would not be
hard to find, but they do not have the broad pumping
bands which make ruby so attractive. Any incoherent
lamp used to pump them is likely to be largely wasted
because only small portions of its output spectrum can
be absorbed by the rare earth ions.

     At TRG, Inc. Gordon Gould and other associates in-
cluding Richard T. Daly had an Air Force contract to
work on optical masers which was, at least in part,
classified. They invited me to visit and give a talk
in the spring of 1960, and we exchanged ideas about work
on spectroscopy of rare earth ions of the sort that
might be useful for optical masers. However, everything
we discussed presented formidable problems and an
operating laser did not seem close.

     Sometime early in 1960, we heard from one of Bell
Labs management that Hughes Aircraft Research Laboratory

in Malibu was also working on optical masers. This point did not register very sharply, because I did not know the people there at all well and did not believe that they had anyone with any optical experience. Nor did I know about the article which was published by Harold Lyons in the Hughes news magazine in which the use of a ruby as a three level maser material was proposed. T.H. Maiman's name was familiar from work he had done on ruby microwave masers, including a paper at the Quantum Electronics Conference.

I spent the spring semester of 1960 at Columbia University as a visiting associate professor. I had been asked to do this because Townes was away at the Institute for Defense Analysis and it seemed good to have someone knowledgeable around in case his students had difficulties between his weekly visits, and also to teach courses. It was not possible for me to move to New York and so I commuted there daily, spending half a day or so at Bell Labs every week. This was an exhausting ordeal and I ended up the semester quite ill with a succession of colds and an infection which took weeks to clear away. But during that spring, I received a paper from Physical Review Letters to referee. In it, T.H. Maiman described some experiments on excitation of ruby by a bright light or flash and made quantitative measurements on the fraction of atoms excited. Maiman's paper indicated that some percent of the atoms could be excited, although it was not possible to tell from the text whether he felt optimistic about being able to produce about ten times greater excitation needed to get more in the excited than in the ground state. I think I suspected that he was interested in optically pumped microwave masers. This manuscript was published in the May 15, 1960 issue of Physical Review Letters.

Late in June, there was a conference on Coherence in Optics at the University of Rochester and some of the Bell Labs people attended. They heard there, from Malcolm Stitch, that Maiman had succeeded in operating a ruby laser. This caused both excitement and puzzlement. Then, in early July there was a press conference at which Hughes announced some information about Maiman's attainment of stimulated emission in ruby. Accompanying the press announcement was a photograph showing Maiman with a rod, very much as I had described, inside a flashlamp. By that time, several people at Bell Labs were working with flashlamps, and as there were not many such lamps on the market, it was easily recognizable as

a General Electric FT-524. Preprints of Maiman's
manuscript for publication were sent out to trade
magazines, and we obtained a copy from a magazine writer who came to Bell Labs to get our reactions and find
out what Bell Labs was doing. There was some skepticism,
but it seemed quite convincing to me. So, many of the
people who were trying to produce various kinds of solid
state optical masers, began to try to obtain laser action in ruby.

Among them were Garrett and Kaiser, with some help
from Walter Bond who was developing techniques for
polishing and coating the crystals, and also D.F. Nelson
and R.J. Collins. These were in different departments
of Bell Labs and physically separated by a considerable
distance in the two different buildings of the Laboratory. A few weeks later, I heard from TRG that they
had gotten their ruby laser to operate the night before.
And so I went to ask Collins and Nelson about their
progress. They felt that they needed better diagnostic
equipment than they had, in particular a better spectrograph to tell what was happening. I had acquired a good
spectrograph for my research and so I joined their effort
and they moved their equipment down to my spectrograph.
Within a day or two they also had achieved laser action.
I remember they were using a General Electric FT-524
lamp which was rated at a maximum input of 4000 volts
from a 400 microfared capacitor. At this rated output,
the optical maser threshold was not achieved so they
raised in input voltage to 4200 volts, 200 volts above
the manufacturer's rating, and their success was
attained. It does not pay to be gentle when you have a
threshold effect! Lasers are nonequilibrium devices and
you sometimes have to be fairly violent to get sufficiently far away from equilibrium.

We did not start out by seeing the beam, but by
looking at an oscillograph showing the output of a
photomultiplier which received the light filtered
through the spectrograph. When laser action was achieved, there was a brief burst of much more intense light
than the fluorescence which was always emitted whenever
you excited the ruby crystal at all. The experiment was
of course quite improvised in many respects and the
bright light from the flashlamp lit up the whole room
and made it very difficult to see what was going on
other than what the instruments told us. In fact, we
were not at all sure that the beam could be seen if
there were a beam. The light from the ruby is at a

wavelength so long that the eye is two hundred times less sensitive than in the green and the pulse lasts only about one two-thousandth of a second.

But the first thing we could do was to find out whether the ruby optical maser had any of the predicted properties. One of these was that the light output should be more monochromatic, that is, should be confined at a narrower band of wavelengths than that of spontaneous emission. It should be directional and it should be coherent, according to the theory of Townes and myself. We set out to try and check these points. The frequency spread could only be measured by flashing the laser repeatedly, photographing the output on an oscilloscope and changing the wavelength setting of the spectrometer between successive flashes. This was something of a problem, because the metal coatings on the ends of the ruby rod, which were usually gold, did not last long at the high intensities of the experiment.

Directionality, which we now can see very easily by just looking at the beam, seemed hard to investigate because we did not know whether we could see the beam. The obvious way to do this was to put a camera focussed for infinity where it could receive the laser output and see whether the laser produced a small spot. A red filter would have to be placed over the camera to screen out the white light from the flashlamp. This was one of the things we were going to get around to but it finally began to be such an obsession with me that after one sleepless night I came in prepared to tell everyone that I was ready to fight to do this experiment next. At this point everyone agreed without a fight, and we found that the output was indeed focussed to a small spot indicating a beam divergence of about a hundredth of a radian, or a half a degree.

Now Maiman had indicated in his paper that he would not expect to get directionality because of reflection from the side surfaces of the crystal. So we deliberately left the sides of our crystal rough in order to minimize reflection from the side walls. Thus I really did expect to get a beam and was gratified when it was obtained. It was a bit crushing a few days later, however, when Garrett, Kaiser and Bond also observed a beam just as good as ours while the sides of their crystal were polished. It was some months later before we realized that the pump light is actually focussed within the crystal so that the degree of excitation is

higher on the axis than it is near the walls. This
means that while amplification is being obtained near
the center of the crystal the outer parts are still
absorbing and so reflections from the side walls are
prevented.

In our photographs, we noticed that the spot had a
grainy substructure and so we found that the output
came in filaments. This helped to explain why the beam
was not even more perfectly directional than it was, as
some diffraction spreading is inevitable from these
small individual filaments. It also showed how laser
action could occur in solid substances that were known
to be highly imperfect. Indeed, study of the linewidths showed the ruby crystal, which we used for our
earliest laser experiments, to be the most badly strained
one I have ever encountered. But somehow there happened
to be some small paths over which light could find its
way from one end mirror to the other and if the gain
was high enough to overcome the diffraction losses in
such a thin column, lasing could occur.

Not very long afterward, Garrett and Kaiser took
the trouble to box in the laser so that the stray light
from the flashlamp could not reach the eye and light
could only come out through a hole at the end of the
ruby rod. They found, and everyone was excited to
realize, that the beam indeed could be very easily seen
as a bright red spot where it struck the wall. Maiman
also apparently observed the directionality about the
same time, for he submitted an abstract to the meeting
of the Optical Society of America and his paper was
presented at their meeting in October 1960.

One day, George Devlin asked "Is there any sign of
hysteresis?" This would not be surprising because many
oscillators do tend to show an overshoot when oscillation
starts or stops, and so we looked more carefully at the
oscilloscope trace. Fortunately, we had acquired a
Tektronics 555 dual beam oscilloscope in which one beam
gives a magnified picture of a small portion of the
trace displayed by the other beam. That is, we could
delay the second beam sweep so that it covered an interval of one hundred microseconds stretching between five
hundred and six hundred microseconds after the initia-
tion of the flash. When we did, we could see that the
output was indeed quite spiky and instead of being a
burst of about five hundred microseconds it was in fact
a series randomly spaced, very intense, one-microsecond
pulses.

To prove the coherence was somewhat more difficult experimentally. We knew that if the light coming out of the laser rod was spatially coherent, we could tell it by doing a Young's two-slit diffraction experiment with the two slits being right in the plane of the laser end mirror. Alternatively, we could have one wide single aperture and observe diffraction maxima and minima in the light coming out through this small opening. The troubles were purely the experimental ones of laying down the suitable patterns on the mirror and then having the mirrors hold together long enough to do the experiment. We had not yet learned how to make dielectric mirrors suitable for ruby laser operations. The single aperture experiment was done first and the results showed definite coherence across the width of the slit which was about 50 micrometers by 150 micrometers.

Between the two groups at Bell Telephone Laboratories we had amassed good experimental proof of the predicted properties of an optical maser. We therefore decided to submit a Letter for publication to Physical Review Letters and we set a cut-off date beyond which we would stop experimenting and finish the manuscript. The single slit diffraction pattern was obtained before the cut-off date but the double-slit was not until a day or two later. The double slit results were therefore submitted separately as a paper for oral presentation at a meeting of the American Physical Society.

In writing this Letter we were concerned, of course, to present our findings clearly and concisely. However, we were afraid to use the title "optical maser," because we had heard some reports that Maiman's letter reporting his very important results had been rejected by Physical Review Letters and we thought that the reason might be because the editors of that distinguished journal had previously expressed a disinterest in further papers on masers as being too concerned with devices for a journal at the frontier of physics. We therefore called the paper "Coherence, Narrowing, Directionality, and Relaxation Oscillations in the Light Emission from Ruby," and it was coauthored by R.J. Collins, D.F. Nelson, W.L. Bond, C.B.G. Garrett, W. Kaiser and myself. I learned much later that the reason for rejection of Maiman's paper was quite different. The editor thought, mistakenly, that it was just a small extension of the paper which he had published very recently, so that it violated their rule against serial publication.

We also were concerned in that paper to be more
specific than Maiman had been in describing just exactly what we had used and had done. We therefore pointed
out that our laser rods had been five millimeters in
diameter and 4.0 cm long even though this was about the
dimensions of the one shown in the newspaper photograph
of Maiman. We also specified that the lamp used was a
General Electric FT-524. It was a year later that I
learned from Donald Buddenhagen that Maiman had in fact
not used the FT-524 in his experiments. It had been
shown in the press photograph because the lamps which
he had actually used were all broken by that time. In
a way that was fortunate, because the diameter of the
FT-524 was larger than the lamp which Maiman used
originally and had enough space in it to permit us to
put in a small Dewar flask and do experiments at lower
temperatures. We also learned much later that Maiman's
original crystal had also not been as long and narrow
as the one shown in the photograph, but that was not
published until some time later.

While this work was in the course of publication,
the Laboratories felt it important to demonstrate that
this work had something to do with communications which
is the prime task of the Bell System. They therefore
arranged to transmit pulses of laser light from the
Holmdel Branch of Bell Telephone Laboratories to Murray
Hill, an airline distance of about twenty-five miles.
There was known to be a direct line of sight between the
two laboratories, and there was a tower at Murray Hill
designed for microwave propagation experiments to and
from Holmdel. The experiment was carried out by Boyle,
Collins and Nelson and they were quickly successful in
not only seeing the beam by eye but photographing an
oscilloscope trace recording in detail the individual
pulsations or spikes which simulated a message that
might later be encoded on a laser beam. All this was
reported at a press conference soon after the publication
of our Physical Review Letter and received widespread
attention.

Meanwhile, I had been continuing my studies of the
dark ruby spectrum hoping to unravel the pair lines and
find out which ones belonged to which kinds of pairs of
chromium ions, nearest neighbors, second nearest and so
on. While that had not gotten very far, I came to
realize that the intensity of these pair lines was
really strikingly high. This indicated to me that they
were being pumped by the much more numerous isolated

atoms. For example, if the concentration is a tenth of a percent, then only one in a thousand of the chromium ions will happen to have a neighbor at the adjacent particular crystal lattice site. Yet, even though the pairs could not be as numerous, the lines could be as strong or even stronger than the lines of the isolated ions. Indeed the absorption was relatively weak, as it should be since paired ions are few, but the emission was strong. This indicated that there was an even more efficient method of exciting the pair lines than I expected and, while I did not want to get into competition with all those who were doing laser work, I finally decided to try out the pair-line ruby laser. By that time, in the fall of 1960, I had acquired a power supply and large flashlamps. The experiments were successful and George Devlin and I found that laser action could be obtained on either of the strongest pair lines separately or simultaneously, with or without laser emission at the R-line. This gave an interesting proof that the 7009Å line and the 7041Å line came from distinct, separate systems and that they are both separate from the R-lines.

We had avoided the word maser in the paper which Bell Laboratory people had sent to Physical Review Letters. But it was apparent by that time, that not everyone had understood that what we were talking about was really coherent stimulated emission from atoms in a resonator. I therefore decided that I would use the term optical maser in this paper, even if that did mean automatic disqualification from Physical Review Letters, and sent it instead to the slightly slower but equally reputable Physical Review. Physical Review Letters, which provides quick publication of important new results, has to be somewhat selective and must reject a substantial fraction of the papers submitted if it is to maintain a rapid publication schedule. Thus they can sometimes apply somewhat arbitrary criteria like the ban on masers which I supposed to be in effect. Indeed they must do so, because the initial discovery in any field is often followed by a growing flood of follow-up papers of gradually diminishing urgency.

However, somewhat to my surprise the paper ended up being published in Physical Review Letters. It happened because on the very same day that the paper by myself and Devlin arrived at the editorial office of the two journals, they also received a paper by I. Wieder and L.R. Sarles reporting stimulated emission from the

pair lines in ruby. The first I knew of it was when the
Wieder-Sarles paper was sent to me to referee. I of
course said that it looked like good work to me. Then
the editors were faced with a necessity of treating both
comparable papers on a similar basis, either both in
Physical Review or both in Physical Review Letters and
they chose the latter. Thus both papers describing laser
action in dark ruby appeared in the February 1, 1961
issue of Physical Review Letters. Though the dark line
ruby laser has been confirmed, it has not been very much
explored because it has so far been operated at temperatures considerably below room temperature.

Around the end of November, 1960 I received a
telephone call from Mirek Stevenson, who had been a
graduate student with Townes at Columbia but was by that
time at the IBM-Watson Research Laboratory. He told me
that he and Peter Sorokin had obtained laser action in
two more substances, calcium fluoride containing divalent samarium ions and calcium fluoride containing
trivalent uranium ions. Both of these materials operated at cryogenic temperatures. Sorokin and Stevenson's
results were already in course of publication in the
IBM Journal and appeared a day or so later. I remember
this event not only because it was one of the earliest
laser materials, probably the second and third laser
materials to operate, but also because a day or so
later when the IBM group announced their results I had
a telephone call from a reporter on the New York Times
asking for comments on it. Fortunately, by that time I
did know about it and I was able to help him get his
story straight. For instance, I suggested that it was
important for him to mention that the samarium and the
uranium ions were respectively divalent and trivalent,
and so it appeared in the New York Times story. At that
time, there were not as many good science writers as
there are now, and I have often been impressed by the
care in which the New York Times took to verify their
stories.

The other memorable aspect was that Stevenson told
me sometime later that he had accepted the challenge of
finding an optical maser material as a management problem. Even as a graduate student he had been very interested in investing and had been extremely successful in
the stock market. Subsequently he founded several companies, mostly in the field of investment and investment
research. But when presented with the problem of quickly constructing an optical maser he went at it by seeking

out the sources of supply for each of the components, that is the crystals, the flashlamp and the power supply and assembling them from these sources so that the experiment could be done in a hurry. This contrasted somewhat with a do-it-yourself approach which most of the others in the field followed. In this case it certainly did pay off, because Garrett and Kaiser at Bell Labs had been working on similar substances, but had not been able to reach the point of getting suitable samples installed in suitable equipment for tests. Undoubtedly, Sorokin also played an important part in this work for he has remained active and has indeed produced several of the most important advances in laser physics since that time.

The first public demonstration of an operating laser was given at the Nerem Electronics Meeting in Boston on November 17, 1960, accompanying my talk. Lewis Winner, who organized the Nerem Meeting had invited me early in the summer, before our experiments on lasers had advanced very far. Fortunately, by the time of the talk our results on the properties of ruby lasers had been published and my colleagues agreed that I could talk on them as well as about my own theoretical work. A large ruby laser power supply was transported to the hall and set up and we demonstrated that a bright red flash could be produced and shown as a momentary bright spot on the screen. We knew of nothing more spectacular to do with it at that time.

A few months later, in January of 1961, George Devlin and I were able to load an optical maser into the back of a station wagon and drive it to Toronto for a similar demonstration at a meeting of the Royal Canadian Institute. But by June of 1961, Bob Ammons had built a really portable ruby laser for Robert J. Collins to take and display at the International Commission for Optics Meeting in London. It used a commercial 200 watt-second photoflash power supply and a small ruby rod with a flashlamp and a little elliptical cylinder reflector.

During the same period, we were finding out some things that these early lasers could do. Willard Boyle showed that if the beam from a ruby laser was focused to a small spot on the surface of an absorber, it would vaporize a bit of that material, producing a white-hot jet. The temperature was so high that even the most refractory materials, such as carbon, could be instantly vaporized. Boyle realized that lasers could then be

used for drilling holes and all sorts of materials proc-
essing. Indeed within the next few years the Bell System, in their Western Electric manufacturing division,
put lasers to use drilling holes in diamonds which were
to be used for wire-drawing dies. Again, Boyle did not
publish his results but I was able to use photographs of
one of these laser produced jets to illustrate in articles which I wrote for Scientific American and the Solid
State Journal. Very soon people began to drill holes
in razor blades by focusing ruby lasers onto them. As
lasers were made larger, it became possible to drill
holes simultaneously through several razor blades
stacked one behind the other. Soon, someone suggested
that laser output power should be measured in Gillettes!

From the beginning, writers of popular accounts in
the newspapers and some people in the military expected,
or at least hoped, that lasers would fulfill the old
dream of a death-ray. We had a good bit of fun with
this notion, which was so very far beyond the capabilities of even the largest lasers. I have often shown a
slide of our "death-ray countermeasures," which I made
at that time. It shows some suits of shining armor, of
the kind that knights used to wear. It was easy to calculate that an unprotected two hundred pound man could
be completely evaporated by about two hundred million
one-joule shots from a typical ruby laser. If we could
deliver them at the rate of one per second, which was
rather better than we could do, he would only have to
stand there for six years. Still, it was obvious even
then that, once the principle was established, you could
expect ultimately to have very large sustained as well
as pulsed powers even though we did not yet know how to
achieve them.

But the problem remained of what one could do with
the small, expensive, heavy pulsed lasers then existing.
I moved to Stanford University in September 1961 and a
few months later, in January of 1962 appeared on a local
television program called "Science in Action." I wanted
to illustrate my remarks with experiments but had only
one very small ruby laser not even big enough to drill
a hole in a razor blade. G. Frank Imbusch, who was then
one of my graduate students, suggested that we could
break a balloon with it and this was tested and found to
be possible. So Imbusch, Linn Mollenauer another student,
and I went to the studio. While I was rehearsing, the
students fixed the balloon with a cardboard base and
fins to look like a rocket ship, and positioned it in

front of the laser. I can assure you that I was much
concerned as to whether the thing would actually work or
not as the program was being broadcast live. Fortunately, it did work and I began showing balloon-breaking
demonstrations. A year or so later, when I was in
Washington I spoke at a meeting of a group of the International Scientific Radio Union (U.R.S.I.) which happened
to meet at the same time as the Optical Society. We
were therefore able to make a demonstration with a large
ruby laser kindly provided by Trion Instruments (later
Laser Systems, Inc.). With this laser we could not only break a blue balloon but we could let the light pass
first through a red balloon which did not absorb the red
light and was unaffected. This made it a more spectacular demonstration and showed that the color of the
light had something to do with it.

Still, the mind works slowly and it was not until
near the end of 1963 when it occurred to me that the
balloon breaking experiment could be done with a dark
blue balloon inside a clear outer balloon. I realized
this as I was sitting with my son at the San Francisco
Zoo and watching other children carrying such balloons.
Of course such a demonstration is a very good illustration of something that could be done with a laser
and not easily in any other way, for the outer balloon
remained unharmed as it did not absorb the light. This
is also a good illustration of the use of lasers to
accomplish things that are at otherwise inaccessible
places, for example as they are used for surgery on the
retina of the eye. To make the stunt more effective Ken
Sherwin, the technician with our research group, built a
small, portable ruby laser into the housing from a toy
ray-gun replacing the flashlight with which the toy had
been sold. I showed this first at the meeting of the
American Association for the Advancement of Science at
Cleveland in December, 1963. It was at that time, while
getting ready for the A.A.A.S. Meeting, that I realized
that laser erasing was possible although my little ray-
gun did not have enough power to erase more than a small
portion of a typewritten character.

The idea of a laser eraser appealed to me, not only
because it was elegant and useful, but also because it
pointed in quite the opposite direction to the thinking
of most of the efforts seeking laser applications. Here
was something that certainly could be done, and for
which there was a very large potential market. The only
problems were engineering and economic. In this it

contrasted sharply with attempts to make laser weapons
for which there were very large amounts of money, but
which no one knew how to do. Naively, I thought that I
could just announce the idea of a laser eraser and
people would start to make them. When this failed, I
was urged to apply for a patent and did. It was apparent that nobody would make the substantial investment
to produce laser erasers without at least the protection
of a patent. The patent was finally granted in 1970,
but it may be that even that will not be sufficient to
overcome the economic obstacles to laser eraser use until suitable lasers are put into quantity production
for some other purpose.

Perhaps a little should be said about the atmosphere surrounding the laser research and the way information was communicated. The initial paper by
Townes and myself setting forth the requirements for an
optical maser and indicating the properties needed for
suitable materials did inspire both considerable interest and considerable skepticism. A number of very good
reasons were advanced why lasers might not ever work,
but some people did take the possibilities seriously
enough to work on them. When lasers were actually
demonstrated in 1960, they did cause considerable excitement both in the popular press and in the scientific
and engineering community. Peter Franken had described
later, at a Symposium of the Optical Society of America
in 1971, the atmosphere at the Optical Society sessions
on lasers in the spring of 1961. "...... I recall now,
just ten years ago on the nose, the first meeting of
any professional society on the laser. That was in the
spring meeting of 1961 in Pittsburgh of this Society and
many of you may not have had the opportunity to be there.
Let me spend a minute and tell you what it was like:
sheer panic. At most meetings, just to give you one
parameter, at most meetings people carry some cameras
and a man will show a slide, that is have a slide shown
at his request and you'll hear a few clicks. At that
meeting every time a slide was projected it was like the
sullen rumble of semi-quieted, semi-automatic fire. In
fact the high point in that meeting occurred during my
good friend Arthur Schawlow's lecture, which was a chalk
talk and he was talking about some of the puzzles and
mechanisms of the ruby laser. I recall a real key point,
I forget the rest of your talk, but the one point I
remember was your saying "We think that there are two
mechanisms operative in the spiking of the ruby laser,"
which was a big puzzle then. He went to the blackboard,

picked up a piece of chalk and wrote down the number 1,
turned away from the blackboard and a dozen cameras
went off. This is the kind of panic that was going on
......." I think Franken must have picked up a bit of
the excitement himself, for he soon borrowed a laser
and then with his associates Hill, Peters and Weinrich
achieved for the first time optical harmonic generation,
thereby ushering in the new and ever growing field of
nonlinear optics.

     Not only were there scientists, but the early
laser meeting attracted a very large number of out-
siders such as engineers in aerospace companies, who
were eager to garner any scrap of knowledge about these
new devices which they might incorporate in systems and
sell to the government. To me, the high point of this
general excitement over the prospects of lasers came at
the meetings sponsored at the Meeting of the Polytechnic
Institute of Brooklyn in New York in March of 1963
where it seemed that every few feet along the corridor
there was someone else asking me some question or other.
But then the excitement subsided as we expected and
people got down to the serious business of learning more
about the operation of lasers, finding new ones and
finding out what they could do. Moreover, the Tower of
Babel effect become more noticeable as individual
specialties within the laser field grew large enough so
that specialists from areas could hardly communicate
with others and indeed did not feel that it was necessary to do so very often.

     What did I learn from all this? Many things, al-
though no golden rule on how to do good research. To
every suggested maxim for guiding research, it is easy
to find a counter example. I have made many mistakes
and have seen my colleagues make mistakes either over-
looking something which later seemed obvious or even
denying it. Yet a good forthright error is often a
stimulating thing as it challenges others to prove you
wrong. Many times new workers in the field seemed
quite foolish and they foundered initially but then
suddenly they were productive and producing their own
original contributions.

     It also appears that there are times when one
should attack a problem head-on, seriously analyzing
the steps needed to attain a desired goal. This was
what Townes and I did in our analysis of the conditions
for an optical maser. At other times it is better to

admit that we do not know enough and just sit back and
add fundamental knowledge, depending on both instinct
and logic to pick out potentially fruitful areas.

I also believe it is a good thing for some of us
should mix pure and applied research committing ourselves fully to neither. Thus Townes could, as he did
in the late 40's, study deeply the processes in the
ammonia molecules as an example of the interaction between molecules and electromagnetic fields. From this
study, he was able to obtain enough knowledge to invent
the maser. But if he had not had the practical inter-
ests in finding out about applications, then that
particular piece of knowledge might not have ever found
its way to the mind of a person interested in generating
electromagnetic waves.

As one looks at the parallel histories of other
new fields of technology such as electronics or avia-
tion, one cannot help realizing that lasers are still
in a very early stage of their development and that
many of the most exciting experiences of discovery are
still to come.

ATOMIC COLLISIONS PHYSICS - ITS INFLUENCE ON TECHNOLOGY AND SOCIETY

Manfred A. Biondi

University of Pittsburgh

Pittsburgh, Pennsylvania

I. INTRODUCTION

The Physics of Atomic Collisions deals with the interactions among atoms, molecules, electrons and photons (pulses of light). Not only are ordinary atoms and molecules involved, but also those containing excess internal energy (excited atoms and molecules), those stripped of one or more electrons (positive ions), and those which have captured a free electron (negative ions). Atomic Collisions together with Atomic Structure comprise the field of Atomic Physics. Although we shall be concerned chiefly with atomic collisions in this essay, the two sub-fields are closely intertwined - atomic structure determines the nature of the atomic collision processes, while the details of the atomic collisions tell us a great deal about atomic structure.

Atomic collision studies apply to matter in either the gas or the plasma (ionized) state: The breadth of the field makes it impractical to detail all of its effects on modern technology and society; therefore we shall use examples drawn from the personal experiences of the author and his colleagues to illustrate how basic studies of atomic collision processes have played a role in the creation of new technology beneficial to society and in shaping policy for national defense and international agreements. As a result, a rather fragmentary view of the overall progress and contributions of atomic collision physics will be given, presented from the viewpoint of participants in particular events. All of the

149

events chosen for discussion occurred in the post-World
War II period, and so this essay deals chiefly with the
more recent history of the field.

The beginnings of atomic collision studies go back
to observations of the behavior of cathode rays (electrons), canal rays (ions), and X-rays (electromagnetic
radiation) in the 1880's and 1890's by Goldstein, Roentgen,
Thomson, Wien. and others. Shortly thereafter, scientists
such as Thomson, Rutherford and Townsend in England and
Langevin in France developed ingenious methods for applying
the electrical and magnetic techniques then available to
exploratory investigations of the collisions of ions and
electrons with atoms.[1] The early theoretical treatments
of atomic collisions were based upon a "billiard ball"
model of atoms, ions, and electrons, with interactions
between particles limited to the classical electromagnetic
forces. (In 1905, Langevin presented a theory of "elastic" ion-atom collisions based on concepts of classical
physics.) There followed studies of collision processes
leading to excitation and ionization of atoms and molecules, of processes leading to electron capture by neutral
molecules and by positive ions, and of the interaction
of light radiation with atoms, molecules and ions.

One of the major triumphs of these early collision
studies was the demonstration of the reality of internal
structure in atoms. Discrete energy states in atoms had
been postulated by Niels Bohr[2] in 1913 in a first step
toward a new, quantum theory of the atom. In 1914 J.
Franck and G. Hertz[3] showed that the measured energy
losses suffered by electrons colliding with atoms could
be exactly correlated with the internal energy changes
produced in the atoms. For this experimental work, they
received the Nobel Prize in Physics in 1925.

The spread of atomic collision studies to many laboratories in Europe, the United States and the Far East
carried through the 1920's and the 1930's, by which time
the field was already a mature one. The chief limitations
in the experimental programs were the comparatively crude
electronic and vacuum technologies then available. Deficiencies in the latter rendered doubtful the results
of many studies owing to the intrusion of impurities in
the measuring systems. Although the development of the
quantum theory during the 1920's had removed the limitations of classical theory in dealing with atomic collision processes, only the simplest atomic systems could
be treated quantitatively using the mathematical techniques then available.

The chief users of the fundamental knowledge obtained from the atomic collision studies were the rapidly expanding electrical, electronics and communications industries. Many classes of light sources, switchgear for electric power transmission, power rectifiers, welders, and high power electronic tubes are plasma devices, and the processes whereby the plasma is formed, emits light, or is extinguished are critical to their operation. As a result, these industries sought to apply the basic atomic collisions information to the development and improvement of plasma devices and processes.

These generalized technological improvements continue today, as is evidenced, for example, by light sources with higher intensities, better color correction and operating efficiencies and by our ability to transmit electrical power more economically using the extra-high voltage systems made possible by higher performance switchgear. This essay will not be concerned with these more conventional applications of atomic collision studies but instead will attempt to show how basic research in this field has led to wholly new technologies and unforeseen applications.

II.   SCIENCE AFTER WORLD WAR II

As noted in the Introduction, the study of atomic collision processes had its beginnings before the turn of the century, essentially as soon as atomic particles were identified and crude techniques became available for their investigation. Activity in the field grew and matured in the 1920's and 1930's and then levelled off as the new frontiers moved to the realm of nuclear physics. Why then, was there a renaissance in this old, well-established field following World War II?

Prior to World War II, the popular conception of a physicist was one who made abtruse discoveries about the nature of the universe but could do nothing of practical value. Then, during the course of the war, teams of physicists and other scientists developed the wide variety of radars and sonars which gave our forces a decisive edge in combat. Concurrently, in a striking demonstration of how a little-known discovery from the forefront of physics - nuclear fission - could be quickly "reduced to practice" in the development of an awesome weapon and an important energy source, other teams of scientists developed first the nuclear reactor and then the atomic bomb. To add to the growing public wonderment at the direct application

of basic science discoveries to practical uses, a few
years later at the Bell Telephone Laboratories physicists
studying fundamental properties of semiconductors invented
the transistor, and the solid-state electronics age was
launched.

Thus, shortly after the close of World War II several
basic ingredients for a renaissance in many fields of
physics research were present. There was strong support
for basic research on the part of the public, industry
and the government, all of whom were convinced of the
practical benefits which would quickly follow. There was
also the availability of new techniques (e.g. microwaves
and ultrasonics from radar and sonar development) which
had not been used previously in research investigations.
In the case of atomic collisions research the newly developed electronics techniques, which were readily adapted
to the precise timing and control of atomic particles,
removed one of the limitations on the progress of experimental investigations. (The other major limitation, that
of adequate vacuum techniques, was soon removed by the
development of an excellent ultra-high vacuum technology[4]
by D. Alpert and his colleagues.) Thus, the academic
centers for atomic collision studies - M.I.T., Berkeley,
Columbia, Johns Hopkins, and Harvard, to name a few -
experienced a major upsurge in activity, the increased
support and new techniques leading to a reexamination
of some of the older, poorly understood phenomena and to
the discovery of new collision processes. The resulting
crop of Ph.D's then made their way to universities and
to government and industrial laboratories, where in many
cases they continued in fundamental research in the field.

Among the industrial laboratories which greatly expanded their basic research programs in the immediate
post-war period were the Bell Telephone Laboratories and
the Westinghouse Research Laboratories, both of which
figure strongly in the research described in later sections. It was to the latter organization that the author,
A. V. Phelps, and G. J. Schulz came, following our graduate work at M.I.T. We were attracted by the prospect
of continuing support of fundamental research in atomic
collisions studies. Higher management at Westinghouse
had been convinced that, as users of fundamental information in the development of new devices for the electrical industry, they had an obligation to contribute to its
acquisition and further that it was good business to do
so, since in 5 to 10 years a given line of basic research
was likely to lead to a major new development. Government
laboratories such as the Bureau of Standards showed a

similar attitude toward the intrinsic value of fundamental research. The universities, which had always had the desire (but rarely adequate means) to support basic research, were now provided with funding, first by the Office of Naval Research and later by its counterparts in the Army and Air Force, by the National Science Foundation, and by the Atomic Energy Commission. Thus, by the 1950's the renaissance in atomic collisions physics was in full swing, with effects on technology and society which we shall now discuss.

III. CONTRIBUTIONS TO SOCIETY

The results of atomic collision studies have not only been applied to the creation of new devices and new technologies, but they have also given us new means for assessing and improving our environment. Of the many possibilities, let us consider three examples of how fundamental studies, motivated by the curiosity of the scientist, have contributed to areas of direct benefit to society.

Lasers

One of the more glamorous recent developments in science has been the invention of the laser (an acronym standing for light amplification by the stimulated emission of radiation). With its achieved or anticipated revolution in industrial processes such as precision welding and drilling of refractory materials, in medical microsurgery techniques, in communications, in ultra-precise distance measurements*, the laser has been the subject of intensive development during the past decade and a half. The first successful laser was the solid-state ruby laser, which was capable of emitting its light energy only in short bursts.[5] The search for a laser which could be made to emit light energy continuously led to the application of results of atomic collision studies to the development of gas-plasma lasers.

All lasers depend for their operation on effectively playing a trick on nature and turning a "normal" situation

---

*Laser beams bounced from the reflectors placed on the Moon by the Apollo missions will be used to measure the extremely slow drift apart of the American and African continents.

upside-down. Atoms and molecules are characterized by
sets of energy levels, something like the rungs on a ladder. At a temperature of absolute zero the atoms or molecules are in the lowest of these energy states, but as
the temperature is increased some are found in the next
higher energy state, fewer in the next and so on. If one
could raise the temperature indefinitely high (to an infinite value), then equal numbers of atoms would populate
each of the energy states.

In certain cases an atom may change spontaneously
from a higher to a lower energy state, with the result
that a short pulse (quantum) of light is emitted. This
is the process which accounts for the light from the gas-
plasma of a fluorescent light or highway arc lamp. However, in order to obtain laser emission of light (stimulated emission) it is necessary to have more atoms in
the higher energy state than in the lower state; that is
we must, in effect, reach temperatures "beyond infinity"
and create a population inversion in the energy states.
It is here that some subtle tricks, gleaned from basic
atomic collision studies, have been employed to achieve
the required population inversion in a series of gas-
plasma lasers.

In the early 1950's A. V. Phelps, then a graduate
student at M.I.T., was invited to spend a summer at Bell
Telephone Laboratories working with J. P. Molnar on studies of metastable atom behavior.[6] (Metastable atoms are
atoms which are in a higher energy state but can not readily return to a lower state by spontaneous emission of
light radiation). Phelps' interest in metastable atom
studies carried over to his postdoctoral work at the
Westinghouse Research Laboratories, where he developed
very precise optical techniques for determining metastable
atom behavior on collisions with ordinary atoms.[7] He
measured the lifetimes of metastable helium atoms in
diffusing through a gas of ordinary helium atoms and found
that the metastables retained their higher energy state
through numerous collisions - thus they were effective
energy reservoirs.

Now, when metastable helium atoms collide with unlike atoms, new processes can occur. If we return to our
ladder analogy for the energy states of atoms, then different atoms are represented by ladders of differing
heights and rung spacings, the top of the ladder corresponding to the energy at which the atom becomes ionized.
The ladder representing neon has a rung just below the
top which lies at almost exactly the same height as the

metastable helium rung; therefore on collision between a helium metastable and a neon atom the energy may be transferred to the neon (excitation transfer) with high probability. From this excited state, the neon atom soon jumps to a lower state with the emission of a light quantum. Since this excitation transfer process leads to population of the upper energy state of neon without first populating the lower states, we have in principle the method of creating the population inversion required for laser action. In practice it is necessary to achieve conditions in a helium-neon gas-plasma such that the helium metastable atoms are produced in sufficient quantities and survive collisions with ordinary helium atoms long enough to have a high probability of transferring their energy to neon atoms.

As a graduate student at Columbia and then in postdoctoral work at Yale, W. R. Bennett had been studying excitation transfer processes in gases. At about the same time A. Javan at the Bell Telephone Laboratories was searching for a practical method of achieving laser action in gases. When Bennett accepted a position at Bell Labs, an active collaboration with Javan ensued. From measurements such as Phelps' helium metastable lifetimes and reasonable estimates of the probability of excitation transfer from metastable helium to neon atoms (no accurate measurements were available), Javan and Bennett concluded that a helium-neon laser might be practical.[8] In collaboration with D. R. Herriott, Javan and Bennett undertook a program of basic measurements which showed that the rate of excitation transfer, though a bit smaller than hoped for, was sufficient for laser action.[9] They then proceeded to build an operating helium-neon laser in December of 1960.

It is an interesting footnote to the history of laser development to observe that even at Bell Laboratories, which have a reputation for unswerving support of basic research, there was an argument in favor of diverting the funds for these basic measurements of excitation transfer to other, more directly applicable research just three months before laser action was demonstrated in helium-neon mixtures. Fortunately, the intrinsic value of the work was recognized, and its support was maintained.

The helium-neon laser was the forerunner of many new classes of gas-plasma lasers which employ quite different schemes to achieve population inversion. While the helium-neon laser, with its modest power output, has been used chiefly in precision optical measuring techniques

(which are based on its precise frequency capabilities),
other gas-plasma laser produce very large amounts of continuous light energy output (in excess of 10,000 watts)
which makes them suitable for use in new industrial technologies. Let us now examine how basic studies of atomic
collisions contributed to the development of the powerful,
infrared emitting nitrogen-carbon dioxide (N_2 - CO_2) laser.

Certain molecules such as CO_2, if set vibrating like
a tuning fork, will radiate infrared light, and the vibration will be reduced. On the other hand when a molecule of N_2 is set vibrating, the fact that both arms of
the tuning fork are identical N atoms prevents radiation,
and the vibration persists throughout many collisions
with other atoms or molecules. The steps which led to
the application of these two observations to the development of the nitrogen-carbon dioxide laser are rather
interesting.

In the early 1900's, Lewis and Lord Rayleigh discovered that when an electric current was passed through
nitrogen gas, a persistent orange-yellow glow of light
was emitted from the ionized gas long after the current
was interrupted. There were a number of speculative
explanations of this "active nitrogen" phenomenon, as it
was called, but none were definitive. Intrigued by the
phenomenon, F. Kaufman and J. R. Kelso at the Aberdeen
Proving Grounds in Maryland, set out to identify the
long-lived energy source which must supply the excitation
of the afterglow light. For this purpose they used their
newly refined fast flow apparatus, in which the gas or
plasma to be studied was introduced at one end of a tube
and made to flow rapidly downstream, where it was analyzed
by various optical and chemical means. They were able
to show by its effect on other molecules that nitrogen
gas ionized by means of a microwave current was evidently
rich in nitrogen molecules which had been excited to the
first of the "tuning fork" vibration energy levels.[11]
The following year K. Dressler at the Bureau of Standards
confirmed their observations by direct optical detection
of the vibrationally excited nitrogen.[12] In addition,
when Kaufman and Kelso added CO_2 or nitrous oxide (N_2O)
molecules to the flowing active nitrogen, they found that
the vibration was strongly "quenched", i.e. damped out.
These results were published in a journal article in 1958.

Kaufman and Kelso considered that this quenching
was most likely the result of vibrational excitation
transfer from the N_2 to a vibration level of CO_2 or N_2O

at almost exactly the same energy (recall the ladder rung
analogy in the helium-neon excitation transfer case),
although they did not state this in their paper. Five
years later J. E. Morgan and H. I. Schiff at McGill University in Canada specifically identified the quenching
process in nitrogen as resulting from a near-resonance
in the vibrational energy transfer to the CO_2 and N_2O
molecules.[13]

The finding of copious amounts of vibrationally excited N_2 in ionized nitrogen gas raises the question of
how it is formed. One usually expects excitation in a
gas carrying an electric current to result from collisions
of electrons with the gas molecules; however, excitation
of vibration by electrons ordinarily is not very likely.
Using the analogy of the tuning fork to represent the
nitrogen molecule, the electron is like a small, very
light pellet* which has only momentary contact with the
tuning fork during a collision and hence is ineffective
in setting it into vibration. The answer to the puzzle
was provided by G. J. Schulz, who was pursuing a program
of electron collision studies at the Westinghouse Research
Laboratories.

In order to study vibration excitation by electrons,
it was necessary to solve the difficult problem of creating
electron beams of very low but well-defined energy. By
refining techniques developed at Westinghouse and at Laval
University in Canada, Schulz obtained the desired electron
beam properties and, as part of his general program, studied collisions of low-evergy electrons with nitrogen.
His interest in this problem had been aroused some years
earlier by a major discrepancy between the measured energy
losses when electrons collided with nitrogen molecules
and the theoretically predicted values.

From his measurements which showed a large probability of exciting vibration in nitrogen, Schulz was led to
postulate and later to confirm that the electrons caused
vibration excitation by an entirely new process - the
formation of a "temporary negative ion" state. This process is highly efficient because now, instead of a brief
collision between electron and N_2, the electron temporarily attaches itself to the nitrogen molecule, effectively
increasing the duration of the collision, and thus permits
the N_2 "tuning fork" to be set into vibration when the

_____

*An electron weighs only 1/50,000 as much as a nitrogen
molecule.

electron departs. These measurements and the explanation
of the vibration excitation of nitrogen by slow electrons
were published by Schulz[14] in 1959.

Following the success in 1960 in creating a helium-
neon laser, the activities in gas-plasma laser development at the Bell Laboratories were greatly expanded.
C. K. N. Patel joined the laboratories in 1961 and for
a time worked in collaboration with Bennett, W. L. Faust
and R. A. McFarlane on studies of processes useful for
attaining laser action in pure gases and in gas mixtures.
In 1964 Patel, Faust and McFarlane reported to the Washington meeting of the American Physical Society their
success in attaining infrared laser emission at a very
low power level in pure CO_2. Following this, Patel began
to consider ways by which a more efficient and powerful
CO_2 laser could be created.

The basic studies described a few paragraphs earlier
had revealed the necessary ingredients for laser action
- efficient generation of a long-lived energy source
(vibrationally excited N_2) and a means to transfer this
energy rapidly to an upper (radiating) state of a molecule (CO_2 or N_2O) to create a population inversion. Patel
recognized that these processes provided him with the
means he sought, and shortly thereafter (in 1964 and 1965)
he demonstrated selective excitation transfer and achieved
laser action in $N_2 - CO_2$ and $N_2 - N_2O$ mixtures.[15,16]

The helium-neon and $N_2 - CO_2$ lasers, whose history
of development has just been described, are but two of
a large and rapidly growing number of gas-plasma lasers
which emit radiation in the ultraviolet, visible, or
infrared portions of the spectrum. Many different schemes
are employed for achieving the population inversion required for laser action. Some lasers are of low-power,
others of high power radiant output; some operate continuously, others in pulsed fashion. In most cases, the
basic understanding of the atomic collision processes
occurring in ionized gases has made major contributions
to their development and improvement.

                    Pollution Detection

As the concern with the quality of our environment
has increased, technology has sought new means to detect
pollution of our atmosphere, both by local and remote
sensing methods. For remote sensing, high power pulsed
lasers - solid-state, liquid, or gas-plasma - offer unique

means of finding where and how much pollution exists in our atmosphere by an optical version of radar called lidar (for light detection and ranging).

To detect particulate matter such as smoke from a blast furnace stack, a short pulse of light at a convenient frequency is beamed from a laser "transmitter" and an optical "receiver" measures the signal bounced back from smoke particles and the time required for the light to make the round trip. Thus one obtains a picture of the denseness, location and size of the smoke cloud, even when it is many miles away.

Some pollutants, such as sulfur dioxide, are clear gases, and the question arises as to how the lidar can be used in their detection. Fortunately, molecules such as sulfur dioxide absorb infra-red radiation at certain sharp frequencies; consequently, the development of infrared lasers tunable in frequency makes it possible to detect and identify different species of pollutant molecules in the atmosphere. Although the lidar development (which is proceeding at laboratories such as the Stanford Research Institute and the AVCO-Everett Laboratories) does not directly involve basic research, we have seen examples of how the lidars' powerful pulsed lasers - solid-state, liquid, gas-plasma, or combinations thereof - are the result of applications of fundamental knowledge of atomic processes.

A second method of detecting atmospheric pollutants involves local sensing of their presence. Here one wishes to detect very tiny amounts of pollutant in the presence of normal atmospheric components; sensitivities of better than 1 part per million are sometimes desirable in tracing some of the more offensive molecules. A powerful new technique for detecting these molecules is under development at locations such as the Airochem Research Laboratories, Ford Research Laboratories, and the Stanford Research Institute. It is based on the phenomenon of chemiluminescence, i.e., the production of light radiation when molecules react with each other.

A bit earlier we mentioned one example of chemiluminescence, the "active nitrogen" afterglow discovered by Lewis and Lord Rayleigh in 1911. The year before Lord Rayleigh had discovered a greenish-yellow "air afterglow" produced by passing electric currents through air samples.[17] Although he had correctly shown that nitric oxide (NO) molecules were required to produce the glow, he mistakenly assumed that ozone (O_3) was the other mol-

ecule in the chemiluminescent reaction. In 1953, M. L.
Spealman and W. H. Rodebush showed that atomic oxygen
(O) rather than O_3 was the reaction partner.[18] In 1958,
while on leave at Cambridge University, F. Kaufman constructed a fast flow system and employed it in the study
of collision reactions involving atomic oxygen. In the
course of this work he made an accurate determination of
the rate of reaction between NO and O to produce the "air
afterglow" and found it to be quite large.[19] The discovery of the rather different chemiluminescence produced
by the reaction of NO and O_3 molecules (a reddish glow)
was made in 1958 by J. C. Greaves and D. Garvin[20] at
Princeton University. The rate of this reaction was determined a few years later at Cambridge University by
J. A. A. Clyne, B. A. Thrush and R. P. Wayne,[21] who had
adopted the fast flow measurement technique. Again the
chemiluminescent reaction was found to occur with high
probability.

Since each chemiluminescent reaction occurs between
specific pairs of atoms or molecules and produces characteristic light frequencies, it is possible to identify
the reacting molecules with some confidence. Using the
very sensitive light detectors (photomultipliers) currently available it is possible to "see" very small concentrations of molecules by the faint chemiluminescent
signals they produce. Thus, if one wishes to find the
concentration of ozone (which acts as a catalyst in the
photochemical generation of smog) in the atmosphere of
Los Angeles, one has only to draw in a sample of the air
into a flow tube, add controlled amounts of nitric oxide
at a point downstream, and observe the light produced by
use of filters and photomultipliers further downstream.
The basic measurements of the chemiluminescent reaction
rates indicate that as little as a few parts per billion
of ozone in the atmosphere are readily detectable, permitting early warning of its presence before it reaches
smog-producing levels. The same detection system may be
used to measure equally small amounts of nitric oxide
pollution of the atmosphere by turning the system around
- that is, adding controlled amounts of ozone or atomic
oxygen at the downstream point. Such pollution detection
devices are currently being marketed.

What started out as a laboratory curiosity - the
long-lived, visible glows associated with the momentary
passage of electricity through atmospheric gases - has
been patiently investigated step - by - step to gain an
understanding of the processes whereby reactive molecules
collide to produce light. Now, as we seek new and better

ways to detect atmospheric pollution, this laboratory
curiosity has been put to practical use. Chemiluminescence is but one example of atomic collision phenomena
which have seen application in devices which assist us
in assessing and even controlling the quality of our environment.

New Sources of Energy

As the world population grows and the desire for
improved living conditions along with it, ever increasing
amounts of energy are required. Unfortunately, this demand for energy is being met by more and more rapid consumption of limited natural resources such as fossil fuels
and has reached the point where energy generation is a
significant source of atmospheric pollution. Although
some alleviation of demand can be achieved, for example,
by full use of recycling techniques (each pound of aluminum produced from ore requires more than 6 kilowatt-
hours of electricity) and by making more efficient those
devices which consume electricity, there are certain energy requirements (e.g. for the manufacture of fertilizers
essential to efficient food production) which must expand
as the population grows.

In the face of this dilemma, the search for new
sources of energy which tap effectively unlimited resources and add minimal pollution to our environment has
taken on an urgent priority. Almost two decades ago,
following the first successful test of the fusion ("hydrogen") bomb, scientists conceived the idea of harnessing
nuclear fusion reactions, which are the energy source of
stars, in an energy producing device called the Controlled
Thermonuclear Reactor (CTR) or simply fusion reactor.[22]

One process of nuclear fusion occurs when two nuclei
of deuterium (heavy hydrogen) collide at such a high speed
that they overcome the repulsion of their like electrical
charges and fuse together to become a new element - the
light isotope of helium. Once the two nuclei fuse, much
more energy is given out than was required to drive them
together and so we have, in principle, an energy producing
scheme. A strong advantage of nuclear fusion over the
more familiar nuclear fission occurring in uranium and
plutonium fueled atomic power plants is that fusion is
a "clean" reaction; that is, neither does it produce
radioactive waste which must be carefully disposed of
nor does it pose a significant radiation hazard in the
event of a reactor accident which leads to venting to the

atmosphere. An even greater advantage is that the fuel
for a fusion reactor - deuterium - occurs with such abun-
dance in the heavy water component of our oceans that it
is practically unlimited in quantity, whereas deposits
of uranium minerals and of fossil fuels apparently can
only satisfy our energy needs for some decades more.

The obstacle to the immediate production of so attractive a device as the fusion reactor is that sufficient
collisions between high speed nuclei for substantial production of energy only occur deep inside stars or in the
violent, millionth of a second duration cataclysm which
is the explosion of a hydrogen bomb. To obtain a controlled release of energy in a fusion reactor, it is necessary to "heat" the deuterium to temperatures greatly
in excess of $100,000,000^{\circ}$F. All solids melt and vaporize
at temperatures less than $10,000^{\circ}$F, so no material container can possibly hold the deuterium in contact with
its walls. However, at million degree temperatures, all
matter is in the plasma (ionized) state, and in this
electrically charged state, the ions and electrons can
be confined by a magnetic field. it then becomes possible, in principle, to contain the deuterium plasma in a
magnetic "bottle" (specially shaped magnetic fields) separated from material walls by an insulating vacuum.

The problem of heating a plasma to these incredibly
high temperatures and stably confining it with strong
magnetic fields became the goal of the plasma physicists
who joined newly formed CTR centers at Princeton, Livermore, Los Alamos and Oak Ridge almost two decades ago.
They had two classes of problems to solve - those that
involved concepts of plasma physics, e.g. stability of
a confined plasma, energy losses by waves in the plasma,
etc. and those that involved atomic collision processes
in plasmas. Belonging to the latter class are questions
of how to avoid overwhelming energy drains from the exceedingly hot plasma, for example, by electron transfer
from the ever-present, cold impurity atoms to the hot
deuterons (ionized deuterium). Such collisions would
neutralize the deuterons and let them escape from the
magnetic "bottle" to the material walls of the CTR device.
In addition, the impurity atoms ionized by the electron
transfer would radiate away large amounts of energy as
they were heated in the plasma. A second class of atomic
collision problems arose in conjunction with the many
novel approaches being considered for the production and
replenishment of the very hot deuterium plasma, their
feasibility resting on the probability of such processes
as ionization by heavy particles and molecular ion dis-

sociation (breakup).

     To solve these problems, the scientists engaged in
fusion reactor development not only made extensive use
of the published literature concerning the basic atomic
collision processes, but they also sought more direct help
from the scientists engaged in atomic collisions research.
Some of us were asked to act as consultants to the various
CTR centers - the author was involved with the Princeton
and Livermore groups - and we spent substantial periods
of time, several weeks or longer, going over in detail
the various approaches to controlled fusion that were
being attempted. In the course of these intensive interactions between the plasma physicists and the atomic physicists, the latter provided critical evaluations of the
available information concerning pertinent atomic collision rates or made estimates where available data were
sketchy.

     A personal example may serve to illustrate this point.
At Princeton, the CTR group under L. Spitzer and M. Gott-
lieb was following the "Stellerator" approach to controlled fusion research, in which a low energy plasma was
generated in a confining magnetic field shaped like a
race-track. By application of suitable electric and
electromagnetic fields at various points around the race
track the ions and electrons in the plasma were accelerated and thus heated. However there was concern that,
at the low energies at which the heating was started,
there would be excessive loss of plasma as a result of
collisions in which the ions recaptured electrons (recombination) to form neutral atoms. N. D'Angelo, a theoretical physicist in the group, carried out some preliminary
calculations which suggested that, as a result of the
strong attraction between the oppositely charged electrons
and ions, such neutralization by recombination would be
exceedingly fast in a plasma.

     In view of the author's previous studies of electron-
ion recombination processes (see Sec. IV) D'Angelo presented his theory for comment. A close examination of
his calculations indicated that, while he had correctly
calculated the large probability that an ion would draw
an electron into its vicinity, he had not provided adequate means for permanent capture of the electron to form
a neutral atom. (The situation is somewhat analogous to
a comet being drawn by gravitational attraction to pass
close to the sun, only to fly back out into space again.)
Only if one provides some means for the electron to lose
energy while it is near the ion will it be captured per-

manently. When this deficiency was pointed out to him,
D'Angelo considered the problem further and found that
the required energy loss was provided by a collision with
another of the plasma electrons while both electrons were
close to the ion. This new recombination process predicted by D'Angelo[23] has proved to be of importance not
only in CTR plasmas but more generally in high density
plasmas encountered in the laboratory and in a variety
of electrical devices.

By such direct interactions between the plasma physicist and the atomic collisions physicist, the controlled
fusion research program received guidance which could not
have been achieved solely by the use of the basic literature. We shall see more of this additional role of scientists engaged in fundamental research - that of advisor
and evaluator - in succeeding sections.

The progress in plasma physics research in the area
relating to controlled fusion has been substantial in the
intervening years, and most of the early problems involving atomic collisions have been solved. While the goal
of achieving a successful fusion reactor seems still to
be at least a decade away, the value to society of such
a nearly inexhaustible, clean energy source makes the
continued long-term effort well worthwhile.

IV.   EFFECTS ON NATIONAL POLICY

In the preceding sections we have seen a few examples
of how applications of basic discoveries contribute to
the development of new devices and technologies. Let us
now consider how the knowledge derived from fundamental
research sometimes plays a role in formulation of policy
decisions by our government. By way of introduction a
bit of scene-setting, I shall describe my own early work
in the field of atomic collisions. This description also
serves to illustrate how a close interaction sometimes
develops between different fields of science in the solution of basic problems.

In 1946, the author joined the group of S. C. Brown
and W. P. Allis at M.I.T. and was given as a thesis topic
the problem of applying microwaves, newly developed for
radar, to the study of collision processes involving
electrons, ions, and atoms. A microwave technique for
the measurement of free electron concentrations in plasma-
afterglows was soon developed, and the new method was
first applied to the simplest cases - the noble gases

helium, neon and argon - for the study of atomic collisions of electrons and ions.

On the basis of previous studies of noble gas
plasmas in the 1920's and 1930's, it was expected that
the most important atomic collision process would be
diffusion of the electrons and ions in the gas. In
helium, the microwave studies showed this to be the case.
However, in neon none of the measurements made sense
when interpreted in terms of diffusion. Finally, when
the process of electron-ion recombination was considered,
the electron loss measurements followed the predicted
behavior, but the measured rate of loss was some 100,000
times too large to fit the accepted theory of the
recombination process.

Following the development of quantum mechanics in
the 1920's, a calculation had been made of the probability of capture of an electron by positive ion. [24]
As noted in the discussion of recombination in Sec. III,
such capture requires that the electron lose energy,
and in high density plasmas collisions with other plasma
electrons provide the energy loss mechanism. However,
in the case of weakly ionized gases (low plasma densities)
such collisions are too infrequent, and the only way
the electron can lose energy is to radiate it away as it
passes close to the ion. The quantum theory had provided a quantitative prediction of the rate of this
radiative recombination and indicated that it was rather
small. Experimental studies of electron-ion recombination
were carried out during the late 1920's and early 1930's
by scientists such as F. L. Mohler at the Bureau of
Standards and C. Kenty at Princeton University. These
studies gave hints that the recombination might be larger
than the theoretical prediction, but the apparent discrepancy was ascribed to experimental errors resulting
from the relatively crude measuring techniques then
available. Thus, it required the microwave technique,
which permitted accurate determinations of the electron
loss, to establish that in low density plasmas electron-
ion recombination occurred enormously faster than
radiative recombination theory predicted it should.

It is rather exciting for a young graduate student
to discover and document such a large discrepancy
between predicted and measured behavior; however, it
led the author to the rash suggestion that the quantum
mechanical theory of the process was in error. It turned
out that the theory for the radiative recombination process was, in fact, correct, but that what was being ob-

served was a quite different recombination process. How
this was established provides an interesting example of
the interplay between different branches of science.

In 1931, J. Kaplan suggested that the faint green
light emission from the ionized layers of the earth's
upper atmosphere (ionosphere) might come from oxygen
atoms produced by the <u>dissociative</u> recombination of
molecular oxygen ions, O_2^+, with electrons; that is, when
an ionized oxygen molecule recaptured an electron, it
broke into two oxygen atoms which were in light-producing,
upper energy levels. Then, in the late 1930's, ground-
based radio probing of the ionosphere suggested a rate of
electron loss some 10,000 times larger than could be
accounted for by radiative recombination. D. R. Bates
and H. S. W. Massey, two British theoretical physicists
interested in ionospheric atomic collision processes
at first concluded that dissociative recombination was not
probable enough to account for this electron loss. However, by 1947, with mounting evidence to support a large
recombination loss of electrons in the ionosphere,
Bates and Massey suggested that the dissociative process
should be considered further.[25]

Thus, when the author and S. C. Brown published the
laboratory measurements of the very large electron-ion
recombination rates [26] in noble gas plasma-afterglows
in 1949, Bates immediately suggested that the dissociative
process was responsible and presented qualitative theoretical arguments to support his contention.[27] There
followed in succeeding years a series of laboratory
tests which established that dissociative recombination
was the extremely efficient electron capture process
observed in the laboratory studies.[28] (The reason for
the 100,000 times greater efficiency of the dissociative
relative to the radiative process is that the energy
which must be lost when the ion captures the electron
can be given to the atoms present in the molecular ion,
causing them to fly apart (dissociate), while in the
radiative process light energy must be generated during
the electron-ion collision.)

The debt owed the ionospheric physicists in suggest-
ing the nature of the laboratory recombination process
was partially repaid a few years later when E. Gerjuoy
and the author published a paper[29] showing how the
laboratory measurements of dissociative recombination
and other relevant atomic collision processes were able
to provide a consistent picture of the behavior of the

E region of the ionosphere, removing some of the contra-
dictions inherent in earlier models. The cooperation
and interchange of ideas between these two branches of
science - ionospheric physics and atomic collision
studies - has developed until now each is of great assistance to the other in advancing the frontiers of our
knowledge. Let us now consider examples of how such
knowledge sometimes figures in the formulation of
national policy.

How Credible the Deterrent?

Following the post-war development of the hydrogen
bomb and the intercontinental range ballistic missile,
it has been national policy to rely heavily on land-based
ICBM's (in conjunction with nuclear powered submarines
and strategic bombers) instead of on large standing
armies for the deterrence of aggressive actions on the
part of other nations. The object has been to establish
a "credible deterrent", a protected force of such missiles
that could survive any "first-strike" and strike back
at a country launching a missile attack on us. As
part of this strategy, there has been simultaneous
research and development on a ballistic missile defense
system to protect our strategic missiles.

Given this background, it is not surprising that
the results of the first high altitude nuclear tests[30]
in 1958 were greeted with dismay by those responsible
for development of the missile defense system. For it
was in these tests that the phenomenon of "Nuclear
Blackout" was first encountered. The term is used to
describe the situation which results when the tremendous
energy from a high-altitude nuclear explosion hits the
upper atmosphere of the earth and ionizes it for distances
of hundreds of miles around. The free electrons in this
ionized region absorb radio and radar waves and so may
"blackout" such systems, i.e. prevent them from operating
properly.

Since the heart of missile defense systems is a set
of radars to detect and then accurately track incoming
missiles to permit interception, the radar blinding
resulting from nuclear blackout could have disastrous
consequences. The factors which determine how badly
and how long the defensive radars are blinded are: (1)
the number of electrons produced in the upper atmosphere,
(2) how frequently they collide with the ambient neutral
molecules, and (3) how long they remain as free electrons.
One way to answer such questions would be to carry out
a whole series of nuclear weapons tests in the atmosphere

and measure the extent of radar blackout in each case.
Not only would such an approach have been extremely
costly, but also the resultant radioactive contamination of the atmosphere would have exacted a price in
increased death rates and genetic defects for generations
to come.

An alternative approach, spurred by a desire to
end atmospheric nuclear testing throughout the world,
was adopted. It involves the calculation of nuclear
blackout effects by using all necessary information
concerning the properties of the upper atmosphere and
the atomic collision processes which determine the
behavior of free electrons - their production, collision
rate and lifetime. Inasmuch as there were gaping holes
in our knowledge of many of these factors, urgent priority was given to acquiring the necessary basic information
in the shortest possible time. It was here that prior
and concurrent work on basic studies of atomic collisions made possible the desired rapid response to
this unforeseen need.

As noted earlier, the author had used microwave
techniques to study electron and ion diffusion and recombination while a graduate student at M.I.T. from
1946 to 1949. On joining the Westinghouse Research
Laboratories he was encouraged to continue basic studies
of electron collision processes in plasma-afterglows
using microwave and other newly developed techniques.
Now, in 1933 N. E. Bradbury had used an electron drift
tube to measure the rate at which slow electrons
attached themselves to neutral oxygen molecules to form
stable negative ions (O_2^-), and in collaboration with F.
Bloch, had given a rough theory for the process.[31]
Inasmuch as the microwave technique offered an excellent
method of observing slow electron processes, in 1950 the
author applied it to the study of electron attachment
in oxygen and obtained a much smaller rate of attachment than had Bradbury.

The capture of an electron by a molecule is somewhat analogous to the case of capture by an ion - energy
must be lost to make the capture permanent. In the
Bloch-Bradbury theory of the attachment process, the
electron is able to stick to the O_2 molecule by setting
it vibrating, thus forming a temporary negative ion.
However, the vibration must be damped by a stabilizing
collision with another molecule if the electron capture

is to be permanent. Bloch and Bradbury had concluded
from Bradbury's experimental measurements that the
necessary stabilizing collisions occurred with great
probability, while the microwave measurements indicated
a much smaller probability.

The reason for the discrepancy was not clear; therefore, the microwave results were not published. When,
a few years later, we were seeking a thesis topic for
L. M. Chanin, the combination of the unsolved discrepancy
and a newly refined drift tube measurement technique
led to a collaboration by Chanin, Phelps and the author
on the electron attachment study. This collaboration
resulted in 1959 in an accurate determination of
electron capture by oxygen molecules[32] which showed
that the actual attachment rate was intermediate between
Bradbury's results and the microwave results. This
new determination was confirmed in independent studies
by G. S. Hurst and T. E. Bortner at the Oak Ridge
National Laboratories.[33]

Since oxygen molecules make up approximately one-
fifth of the upper atmosphere at the altitudes where
nuclear blackout occurs, the lifetime of the free
electrons, which determines the duration of the radar
blinding, depends directly on the rate at which oxygen
molecules capture electrons. Thus, these laboratory
studies occurred at almost the moment that the urgent
need for their results became evident, yet the motivation
was simply a wish to understand a puzzling discrepancy
between two measurements of a basic process.

While the attachment of slow electrons to oxygen
molecules is the process which most critically affects
electron lifetimes in nuclear blackout, for a shorter
time, recombination of electrons with positive ions also
has a major influence. As we have seen, the rate of such
recombination depends on the nature of the electron
capture process. The microwave studies of recombination
of electrons and ions which were first applied to the
simple noble gas plasma-afterglows had established the
very short free electron lifetimes that might result
from the efficient <u>dissociative</u> capture process. Further,
microwaves provided a general technique for the determination of electron capture rates by various molecular
ions. Consequently as a next step, studies of the recombination of electrons with the ions expected in the
upper atmosphere - those of molecular oxygen, nitrogen
and nitric oxide (O_2^+, N_2^+ and NO^+) - were attempted at a

number of laboratories, inasmuch as these were of
interest to ionospheric physicists. These first attempts
were not very successful, since in atmospheric gases one
could not be sure of the precise identity of the molecular ion responsible for electron capture.

The need for these atmospheric ion recombination
rates in nuclear blackout calculations led to increased
support of recombination measurements through the various
research offices of the Department of Defense. Once
again, the availability of suitable techniques developed
in basic investigations permitted a rapid response to
this need. The microwave method of measuring electron
loss was supplemented by mass spectrometric identification
of the ions undergoing recombination to provide quantitative determinations of electron capture by such ions
as N_2^+, O_2^+ and NO^+.[34-37] At present, much of the
desired electron lifetime information has been obtained,
benefiting not only the blackout applications area
but also the ionospheric physics community.

A second critical factor in the evaluation of nuclear blackout effects is the rate at which slow electrons collide with atoms and molecules in the upper
atmosphere, since this collision rate determines the absorption of radio and radar waves. During the 1910's
and 1920's extensive studies of electron collisions had
been carried out using electron beam and electron drift
tubes. Some of the measurements, such as those of C.
Ramsauer[38], were remarkably accurate; however, the available techniques did not permit studies of sufficiently
slow electrons to match the conditions in the upper
atmosphere. Since microwaves permit the study of arbitrarily slow electrons, in the early 1950's Phelps,
Fundingsland and Brown[39] at M.I.T. sought to extend our
basic knowledge by developing a technique to measure
the rate of collision of slow electrons with various
gases. They applied this new method to studies of
electron collisions with various atoms and molecules,
among them nitrogen molecules. Inasmuch as nitrogen
molecules are the principal constituent (about 80%) of
the earth's atmosphere at the altitudes of interest for
blackout, an accurate determination of this collision
rate and its dependence on electron energy was essential
to blackout calculations, and the results were soon applied in calculations of radar absorption.

Now, while the electron attachment, recombination
and collision rates are a central part of the blackout

problem, they are analogous to the visible part of the iceberg, with a large, underlying structure of atomic collision processes which determine the nature of the ions, atoms and molecules which the electrons encounter. The acquisition of the broad spectrum of atomic collision information relevant to blackout is the continuing goal of efforts by a significant part of the experimental and theoretical atomic collision physics and ionospheric physics communities at universities, industrial and government laboratories. The particular examples cited, while presenting only a microscopic view of the overall contributions, show rather clearly how the availability of basic information and new techniques gave a head-start of several years in the evaluation of blackout effects.

The continued acquisition and use of atomic collision information has permitted a more sophisticated understanding of nuclear blackout and related problems in defense against ballistic missiles. This understanding in turn provides a more realistic assessment of the capabilities (or vulnerability) of a given ABM system at any moment and should therefore help prevent responses to imagined threats to the system. This role of basic science in contributing to accurate assessments of technical situations not only can play a role in shaping national defense policy, but, as we shall now see, also can aid the progress of international agreements.

Broadening the Nuclear Test Ban Treaty

In July, 1963, following long and tedious negotiations, an agreement was reached banning nuclear testing in space, in the atmosphere and underwater. To date, 102 nations, including the United States and the U.S.S.R. have ratified the treaty, which represents a significant first step in the broader quest for a workable disarmament agreement. The cessation of atmospheric testing by the major powers (with the exception of France and China) has also led to a remarkable reduction in the radioactive contamination of our atmosphere.

This partial test ban agreement was reached only after the U.S. and the U.S.S.R. were convinced that they could detect clandestine testing on the part of another nation; thus, it specifically excluded underground testing, since at the time seismic detection systems were not considered sensitive and fool-proof enough to identify a low yield weapon test in the presence of normal earthquake seismic wave background. With the development of

sophisticated seismic detection systems, confidence in
the ability to detect and locate underground tests of
low-yield weapons rose, and consideration was given to
broadening the test ban treaty to include underground
testing.[40]

With the prospect of a total ban on weapons testing
came the concern that some nation developing a new
weapons system might attempt to test it beyond the range
of our detection systems - for example, out in space - and
so gain an unsuspected advantage. One of the methods
by which we detect nuclear tests in the atmosphere or
in space near the earth is by keeping a watch for the air
fluorescence "signature" which is produced by a nuclear
explosion in these environments. These detection
systems were initially developed at the Los Alamos
Scientific Laboratories in the group under the direction
of H. Hoerlin and at the Atomic Weapons Research Establishment in England in the group of R. Wilson. These
systems make use of optical sensors which detect the
characteristic pattern of light emitted from the higher
regions of the earth's atmosphere by atoms and molecules
excited by the nuclear detonation.[41]

The energy release of a nuclear explosion appears
in many forms, one of which is a very intense burst of
X-rays. If the explosion is out in space, these X-rays
travel without attenuation until some of them encounter
the upper reaches of the earth's atmosphere and are
absorbed. In this absorption process extremely energetic
electrons are released which in turn collide with the
atoms and molecules of the atmosphere to cause ionization
(and excitation) which produces still more electrons of
somewhat lower energies. This snowballing growth of
electrons is accompanied by excitation of the molecules they strike, and so one has a characteristic
pattern of light radiation, both as to wavelengths
(colors) emitted and as to variation of light intensity
with time as the excitation spreads across the sky. The
optical detection network then recognizes this pattern or
"signature" as produced by a nuclear explosion and is not
confused by other atmospheric radiations such as lightning strokes.[40]

Since the air fluorescence signature becomes weaker
the farther out in space one detonates a nuclear explosion,
willingness to seek a total test ban agreement hinged in
part on confidence that our detection system was sensitive enough to spot any such attempts at cheating. Consequently, as a part of its responsibility under Project

VELA - the nuclear test detection program - the Advanced
Research Projects Agency in 1964 approached T. M. Donahue, an ionospheric physicists at the University of
Pittsburgh, and asked that he arrange a Summer Study
for the purpose of providing a definitive assessment of
the capabilities of the air fluoresence detection
systems.

Donahue's interest and researches in upper atmosphere physics had started with his graduate work at Johns
Hopkins University from 1942 to 1947, and since that time
he and his students have been engaged in ionospheric
calculations and observations programs, as well as
related laboratory studies of atomic collisions. Thus,
he was well acquainted with both the scientists engaged
in upper atmosphere research and in atomic collision
studies. In the summer of 1964 he brough together a
group of 60 scientists and engineers to evaluate the air
fluoresence detection systems.

Some of the group were the designers and developers
of the system components who could provide detailed engineering data on performance, others of us had backgrounds
in theoretical and experimental investigations of atomic
collision processes involving atmospheric constituents,
while a third group had detailed knowledge of upper
atmosphere behavior. The basic research scientists were
drawn from centers of ionospheric studies such as the
Kitt Peak National Observatory, Naval Research Laboratories, Air Force Cambridge Research Laboratories, Bureau
of Standars at Boulder, Johns Hopkins University, University of Colorado and from centers of atomic collisions
research such as the Bureau of Standars, Joint Institute
for Laboratory Astrophysics, Westinghouse Research
Laboratories, and the University of Pittsburgh.

The work of the Summer Study group was carried out
in two two-week sessions, during which time an intensive
examination was made of the basic physical processes
responsible for the production of air fluoresence, of
the various existing optical systems for detecting the
fluorescence, and of possible new fluorescence effects
useful for detection. This work was accomplished by
dividing into small working groups which concentrated on
particular aspects of a problem (e.g. efficiency of light
production), the makeup of each group being determined
by the specialized knowledge of its members. After the
small working groups had reached their conclusions, the

membership met as a whole to integrate the results into
the overall evaluation.

An example of one of the important questions affecting detection sensitivity is the "quantum efficiency"
(the ratio of the energy output in light fluorescence to
the energy input from the X-rays) for production of the
characteristic ultra-violet light emitted by excited nitrogen molecular ions. This light is suitable for the
fluorescence detection system, since it is readily excited, and it occurs in a part of the spectrum where radiation detectors are very sensitive. To determine the
"quantum efficiency" it is necessary to find the probability that an energetic electron will, in one step,
ionize and excite a nitrogen molecule. At present, even
with the aid of large computers, we are able to calculate
ionization and excitation only for the more complicated
atoms and can not as yet obtain quantitative results for
molecules such as nitrogen. Thus, appeal is made to experimental measurements for the necessary information.

There have been three types of measurements which
relate to the determination of the quantum efficiency for
ultraviolet light production: (a) laboratory measurements
of the probability of excitation of nitrogen molecules
as a function of electron energy, (b) laboratory measurements of the light emitted when very fast electrons are
"stopped" in nitrogen gas, and (c) measurements of the
ultra-violet light produced in the atmosphere by nuclear
weapons tests. Unfortunately, the results from these
widely different types of measurements were in conflict;
therefore, a working group was convened to examine in
detail each method of measurement and the interpretation
of its results. After extensive study and some spirited
arguments between members of the group, deficiencies in
the various measurements were identified and taken into
account, with the result that a more accurate value for
the "quantum efficiency" of ultraviolet light production
emerged.

After a recess of 10 weeks to reflect on the results
obtained during the first part of the Summer Study, the
members met again to draw final conclusions concerning
air fluorescence detection. The Summer Study more than
met its objectives - not only did it provide a critical
evaluation of the sensitivity of the existing system -
it also yielded suggestions for improvements in the detection system which extended its effective range and
reduced the number of stations required in the network.
In addition, as a result of the exploration of all facets

of the air fluorescence problem, Donahue was led to suggest a completely new detection scheme of even greater sensitivity.

There have been a number of group studies of this type which have pooled the knowledge of basic scientists (who were led to acquire this knowledge for its own sake and not for forseeable applications) in the solution of a problem of national interest. The active participation by these scientists in the evaluations has permitted the full scope of their knowledge to be used and generally has resulted in an accurate, objective appraisal of the situation under consideration. In the present case, while the hoped-for broadening of the test-ban agreement has not as yet taken place, the contribution of basic science in establishing confidence limits on and improving our detection systems remains, so that when other obstacles are removed, negotiations may once again move ahead.

V. CONCLUSIONS

The foregoing sections have presented a highly personal account of some segments of the progress in Atomic Collisions Physics and the attendant effects on technology and society. These segments were drawn from the author's own experiences and those of his immediate circle of colleagues at M.I.T., the Westinghouse Research Laboratories and the University of Pittsburgh. By intent, the coverage of the field has not been encyclopedic, since a recitation of all new developments would have been overly long and dull, Instead, it is hoped that the examples chosen demonstrate not only the well-known impact of basic research on technology, in which the discovery of A leads to the invention of B, but also the basis which scientific knowledge provides for accurate value judgments affecting our society, as in contributing to decision making at the national level.

In case of inventions which make use of the results of fundamental research, the examples serve to illustrate the fact that the path between discovery A and invention B is only rarely the direct one. Instead, invention usually follows when enough basic knowledge has been acquired to formulate a "working hypothesis", which is often a shrewd extrapolation of what has been determined to an estimate of what is needed in a working device - the laser developments fit this category rather well. Another point brought out by the examples is that basic research at institution X often is put to use in a new

device or process developed at institution Y. While this
may lead to short-term repercussions if, for example, X
represents the Westinghouse Research Laboratories and Y
the General Electric Corporation, the scales tend to balance out, with many institutions benefitting from the
general fundamental discoveries. It would be a mistake
to attempt to channel any substantial portion of basic
research into directions of known applications, since
this almost certainly would cut off the discovery of the
truly unforeseen, and it is these areas which open up
whole new technologies.

The reader may have noted that in the discussions
of developments in the laser and pollution detection
fields, the time lapse between the basic discovery and
its application in a new device or process has become
remarkably short. In a field such as atomic collisions,
we are applying new knowledge almost as fast as it is
being generated. There are some who argue that this
headlong development of new technology is at the root of
society's troubles today, and they yearn for a return to
the simpler "good old days".

In answer to this argument, the new knowledge obtained from basic research is not the source of the difficulty, since knowledge is, of itself, neither good nor
bad. Problems arise, however, when technological developments are embarked upon with too little reflection on
their ultimate costs to society. In these cases it is
the social responsibility of the scientist who sees such
inappropriate applications of research efforts to speak
out and alert society to his concerns. Such social responsibility is more and more evidencing itself in the
actions of organizations of scientists and engineers in
taking stands on technical matters affecting the public
welfare.

We can not go backwards to the simpler society, since
the present world population could not be sustained by
the science and technology of the "good old days". In
fact, even modern technology falls short of providing an
adequate mode of living for a majority of the earth's
inhabitants. Thus, we are very dependent on continuing
advances in basic research and on the applications that
spring therefrom. Society's problem is to develop a
sufficiently mature set of social and political institutions to make the wisest use of the newly available technologies in achieving society's goals.

NOTES AND REFERENCES

1. A detailed description of the early work in the field
    has been given by L. B. Loeb, Fundamental Processes
    of Electrical Discharges in Gases, (John Wiley and
    Sons, New York, 1939).

2. N. Bohr, Philosophical Magazine, Vol. 26, p. 1 (1913).

3. J. Franck and G. Hertz, Deutsches Physikalische
    Gesellschaft Verrichtung, Vol. 16, p. 457 (1914).

4. See, for example, D. Alpert, Handbuch der Physik,
    Vol. 12, p. 609 (1958).

5. For a review of laser development up to 1962, see
    A. L. Schawlow, Scientific American, Vol. 204, p.
    52 (June 1961) and W. R. Bennett, Applied Optics
    Supplement, Vol. 1, p. 24 (1962).

6. A. V. Phelps and J. P. Molnar, Physical Review, Vol.
    89, p. 1202 (1953).

7. A. V. Phelps, Physical Review, Vol. 99, p. 1307
    (1955).

8. A. Javan, Physical Review Letters, Vol. 3, p. 87
    (1959).

9. A. Javan, W. R. Bennett, D. R. Herriott, Physical
    Review Letters, Vol. 6, p. 106 (1961).

10. R. J. Strutt (Lord Rayleigh), Proceedings of the
    Royal Society, Series A, Vol. 86, p. 56 (1911).

11. F. Kaufman and J. R. Kelso, Journal of Chemical
    Physics, Vol. 28, p. 510 (1958).

12. K. Dressler, Journal of Chemical Physics, Vol. 30,
    p. 1621 (1959).

13. J. E. Morgan and H. I. Schiff, Canadian Journal of
    Chemistry, Vol. 41, p. 903 (1963).

14. G. J. Schulz, Physical Review, Vol. 116, p. 1051
    (1959).

15. C. K. N. Patel, Physical Review Letters, Vol. 13,
    p. 617 (1964).

16. C. K. N. Patel, Applied Physics Letters, Vol. 6, p. 12 (1965).

17. R. J. Strutt (Lord Rayleigh), Proceedings of the Physical Society, Vol. 23, pp. 66 and 147 (1910).

18. M. L. Spealman and W. H. Rodebush, Journal of the American Chemical Society, Vol. 57, p. 1474 (1935).

19. F. Kaufman, Proceedings of the Royal Society, Vol. A247, p. 123 (1958).

20. J. C. Greaves and D. Garvin, Journal of Chemical Physics, Vol. 30, p. 348 (1959).

21. M. A. A. Clyne, B. A. Thrush and R. P. Wayne, Transactions of the Faraday Society, Vol. 60, p. 359 (1964).

22. A review of early progress in controlled thermonuclear research has been given by R. F. Post, Reviews of Modern Physics, Vol. 28, p. 338 (1956).

23. N. D'Angelo, Physical Review, Vol. 121, p. 505 (1961).

24. J. R. Oppenheimer, Zeitschrift fur Physik, Vol. 55, p. 725 (1929).

25. D. R. Bates and H. S. W. Massey, Proceedings of the Royal Society, Vol. A187, p. 261 (1946) and Vol. A192, p. 1 (1947).

26. M. A. Biondi and S. C. Brown, Physical Review, Vol. 75, p. 1700 (1949) and Vol. 76, p. 1697 (1949).

27. D. R. Bates, Physical Review, Vol. 77, p. 718 and Vol. 78, p. 492 (1950).

28. See, J. N. Bardsley and M. A. Biondi, "Dissociative Recombination" in Advances in Atomic and Molecular Physics, Vol. 6, D. R. Bates, Ed. (Academic Press, New York, 1970).

29. E. Gerjuoy and M. A. Biondi, Journal of Geophysical Research, Vol. 58, p. 295 (1953).

30. See, The Effects of Nuclear Weapons, S. Gladstone, Ed. (U. S. Atomic Energy Commission, April 1962), Ch. 10.

31. F. Bloch and N. E. Bradbury, Physical Review, Vol. 48, p. 689 (1935).

32. L. M. Chanin, A. V. Phelps and M. A. Biondi, Physical Review Letters, Vol. 2, p. 344 (1959) and Physical Review, Vol. 128, p. 219 (1962).

33. G. S. Hurst and T. E. Bortner, Physical Review, Vol. 114, p. 116 (1959).

34. W. H. Kasner and M. A. Biondi, Physical Review, Vol. 137, p. A317 (1965) and Vol. 174, p. 139 (1968).

35. R. C. Gunton and T. M. Shaw, Physical Review, Vol. 140, p. A756 (1965).

36. F. J. Mehr and M. A. Biondi, Physical Review, Vol. 181, p. 264 (1969).

37. C. S. Weller and M. A. Biondi, Physical Review, Vol. 172, p. 198 (1968).

38. C. Ramsauer, Annalen der Physik, Vol. 64, p. 513 (1921).

39. A. V. Phelps, O. T. Fundingsland, and S. C. Brown, Physical Review, Vol. 84, p. 559 (1951).

40. For technical background information, see the "Nuclear Test Detection Issue" of the Proceedings of the IEEE, Vol. 53, pp. 1814-2097, (Dec. 1965).

41. Testimony concerning the performance of the test detection systems was presented in 1963 at hearings before the Joint Committee on Atomic Energy, 88th Congress of the United States and is published under the title Developments in Technical Capabilities for Detecting and Identifying Nuclear Weapons Tests (U. S. Government Printing Office, Washington, D. C., 1963).

EDWARD TELLER

THERMONUCLEAR ENERGY

Edward Teller

University of California

Berkeley, California

"It will be the first time that man can make direct use of nuclear energy." It was the summer of 1939. Albert Einstein signed a letter addressed to Franklin D. Roosevelt. The only comment he made about the contents of the letter was the first sentence of this paragraph. It was a peculiar statement; I have not forgotten it, though I could only vaguely feel that this statement had the qualities of a prophecy: foreshadowing the future, but leaving the coming events uncertain and unexplained.

I had driven my friend, Leo Szilard, from New York to the far end of Long Island. I was not needed for the discussion. The letter lacked nothing except Einstein's signature. But while Szilard could accomplish many things that no one else would undertake, he could not do what everybody else practiced. He could not drive a car. So I acted as Szilard's chauffeur. Einstein lived in a cottage and Szilard could not give me directions. But on the quiet Sunday afternoon we found a young girl and she knew where Einstein lived. Einstein was famous enough and it was impossible to forget his face.

He received us in slippers, poured some tea, read the letter and muttered: "It will be the first time...". We did not know, but we suspected what would happen to Poland in a few weeks. We had an idea what a nuclear explosion would be like. But we clearly understood what Einstein was saying. All usable energy has come to us heretofore from the sun. The warmth of a summer day, the storms, the coal burning and driving an engine, all

derive their energy from the sun. But the sun itself
draws on an ancient store of nuclear energy. And now we
were on the road toward the direct exploitation of that
basic source, by human hands.

Yet there was something that did not quite fit in
Einstein's remark. What we stumbled upon a few months
ago was the fact of fission. Heavy nuclei can split and
release great amounts of energy. The sun keeps radiating
for a different reason. This reason had been clarified
little more than a year before we drank tea on Long
Island. (It is strange to recall how closely the discoveries concerning the tiny nucleus of the atom followed
each other.) The smallest and lightest nuclei can combine due to the attraction of all the building blocks
of nuclei. This "fusion" likewise releases energy and
furnishes the source of the solar radiation.

That the smallest (and heaviest) parts of the atoms,
the nuclei, should be particularly rich in energy can
be readily understood. Close packing of the constituents
requires great binding energies. Actually, one of the
basic facts of atomic theory is that accurate localization
of a particle is unavoidably connected with violent
motion and great amounts of kinetic energy. The binding
energy has to over-compensate this energy of movement.
In fact, nuclear energies surpass chemical energies
(connected with comparatively loose arrangements of
electrons in atoms and nuclei) a million-fold.

It is more remarkable that these huge amounts of
energy remain untapped under normal conditions of human
experience. All nuclei carry a positive charge. They
repel each other. The agitation due to a high temperature is required to bring the nuclei close enough to
touch and to react.. In the hot center of the sun this
is what actually happens, at an exceedingly slow rate.
Billions of years are required for the sun to exhaust
its nuclear fuel. Here on earth nuclear energy is in
cold storage to be used when appropriate conditions of
a rather extreme nature are generated.

How to bring about these extreme conditions, none of
us could guess at that Long Island tea party. We were
talking about fission, the breaking apart of almost un-
naturally overgrown nuclei. The more widespread, more
normal process of fusion was not more than an obliquely
stated program for future experimentation and engineering.

The first step in obtaining mastery of nature is, as a rule, to understand what we may later influence. The reason for the attempt is not an intention of a useful application. It is rather an intellectual challenge or, stated in a more modest and appropriate manner, simple curiosity.

Why does the sun continue to radiate? Any store of chemical energy would have been exhausted in a few millennia. The obvious use of gravitational energy may have lasted for a few millions of years. But the geological record extends into the past for a thousand times longer period. Only the atomic nucleus seemed capable of furnishing the required source.

It is possible to give a qualitative, but strong, argument in support of the nuclear nature of stellar fuels. We can estimate the temperature in the center of stars. These temperatures happen to be right for overcoming the mutual repulsion of nuclei and for permitting nuclear reactions to proceed at the required speeds.

The simplest way to obtain an approximate value for the temperature of a quiescent star is to set the average kinetic energy, which represents the temperature, equal to one half of the potential energy. The latter is known as soon as a rough distribution of matter within the star can be reasonably assumed. The estimate is made more difficult because there is a strong concentration of matter near the center of the star, giving rise to high gravitational potential energy. The central density is a hundred times the density of water (and also a hundred times the average solar density). But a value of 30 million degrees Fahrenheit can be obtained for the solar center without too much trouble. Furthermore, this temperature, give or take a factor two, is typical of many stars.

The effect of such a high temperature on nuclear reactions is easily calculated. One first has to remember that at a given temperature particles do not have a given velocity but rather a range of velocities. High velocities do occur with small probabilities.

At the same time, the probability for a nucleus to make contact with another nucleus becomes greater if the velocity is great, but becomes small if the nucleus is more nearly charged. Thus hydrogen, whose nucleus carries the least charge, has the best possibility to react,

particularly if the velocities are high. This implies
that we are particularly interested in the few nuclei
of the greatest speed which will be, of course, found
in greater numbers at high temperatures.

It is worthwhile to mention that contact between
nuclei does not occur according to the easily visualized
rules of classical mechanics. One should expect that
if the energy suffices and the collision partners approach head-on the nuclei will touch. This is correct.
But one also should expect that if the energy is insuf-
ficient, contact can never be established. This is not
true and the fact that it is not true is more significant. In atomic physics particles do not always act as
particles, they also act as waves. A particle may be
turned back in its orbit at a sharply defined position.
A wave process does not stop abruptly. It gradually
fades away. In consequence, collisions which in the
classical picture have no chance of an approach close
enough for a nuclear reaction actually do have a small
chance to lead to such a reaction, though the probability declines rapidly as we consider collisions of
smaller energies.

In calculating the probability of nuclear reactions,
two factors must be taken into account. First, high
energy collisions are improbable. Secondly, low energy
encounters will hardly ever lead to nuclear reaction,
but the probability of a reaction increases rapidly with
increasing energy. Actually, the reactions occur mostly
as a result of a compromise: the energy of the effective collisions is high but not too high. This energy
is higher than the average energy of thermal motion
(which is, of course, proportional to the temperature T)
and increases as $T^{2/3}$. The rate of thermonuclear
reactions depends on the temperature as

$$\exp\left[\frac{-\text{Constant}}{T^{1/3}}\right] \quad .$$

All nuclear reactions become slow and unobservable at
low temperatures. But at high temperatures their rate
depends on the constant which appears in the exponent.

Everything that has been said so far applies equally to reactions occuring in the stars and the similar
reactions for which man might create the proper conditions. But at this point an important difference
occurs. Most stars, and in particular the sun, rely on
two reactions. As we shall see, neither of these lend

themselves for the purpose of a man-made energy source.

The sun contains mainly the most abundant type of
hydrogen nuclei, the protons which carry a minimal
charge and approach each other with relative ease. But
if they collide, they do not stick together. That they
react at all is due to the fact that at the moment of
collision one of the protons may turn into a neutron,
which is a building block found in most nuclei. The
proton and the neutron then stay together, having formed
a deuteron, the simplest of all complex nuclei. (All
other nuclei are built, like the deuteron, from protons
and neutrons.)

It is easy to state but hard to imagine how improbable it is that two protons should form a stable deuteron
even if the protons get sufficiently close to each other.
It happens in one of the 10^{24} contacts; the probability
is a trillionth part of a trillionth. A proton will
approach another proton about 10^{11} times (hundred billion
times) per second. Only a small fraction of these approaches results in a penetration of the wave-process
to really close quarters, but even that occurs a few
million times in each second for each proton. Yet in
the lifetime of the sun, which is five billion years,
only a few percent of the protons have reacted.

Incidentally, the deuteron is not the end of the
road. More reactions follow at a comparatively fast
rate until in the end result, out of four originally
independent protons, a strongly bound helium nucleus
(or α-particle) is made which consists of two protons
and two more which have metamorphosed into neutrons.

Still another series of nuclear reactions occur in
the sun, probably to a lesser degree. Heavier and more
strongly charged nuclei, those found in carbon, nitrogen
and oxygen capture protons. This is accomplished without
the need for the proton to turn into a neutron in flight.
That transformation (which is a β-decay) follows in some
cases after the proton has been caught. In the end, an
α-particle is split off and the net result is the same
as in the reaction that started by the interaction of
two protons. Because these higher nuclei repel protons
more strongly, the reaction again takes place at an
exceedingly slow rate. In bigger stars with higher
central temperatures, this so-called carbon-cycle actually predominates. For man-made reactions, however,
this reaction is again impractical.

Of all this I was aware when I listened to Einstein's
laconic reaction. I knew that for the sun thermonuclear
reactions were appropriate. It was less than a half-
born hunch that one could reproduce on earth the stellar
spark. We had at that time something more important
to do. The catastrophe of the Second World War, the
opportunity of nuclear energy appeared too close to allow
much speculation.

In a little more than two years the situation
changed. Roosevelt read the letter after the fall of
Poland. Even before Pearl Harbor, the scientific community became active. Necessity turned the most skill-
ful hands and the most fertile brains toward concrete
applications of science.

I worked in a group at Columbia University which
worked on nuclear fission under the direction of Enrico
Fermi. One day in the fall as we returned from lunch,
Fermi raised the question of thermonuclear reactions.
A nuclear explosion should produce for a brief instant
temperatures similar to those prevailing in the center
of the sun. Could one not use a fission explosion as
a match to start a thermonuclear fire? I objected: the
time available is too short. But, Fermi said, we can
provide much better fuel than is available in the sun.
We could use deuterons, the nuclei of heavy hydrogen.
They approach each other almost as easily as common
hydrogen nuclei. And, when they touch, they react and
produce energy with a high probability. This gains
for us a factor 10^{24}, a trillion times a trillion.

In a few days I told Fermi of the difficulties I
found. They seemed insuperable and for the time being,
we forgot about the possibility. But a few months later
I decided to write down in detail a complete proof that
thermonuclear reactions cannot be used as an energy
source on earth. The more I tried, the more loopholes
I found. At that time, Oppenheimer formed a new group
to consider specific problems connected with nuclear
explosions. When I was asked to join I brought along
my problem. All of us found it exciting. In a few
weeks we convinced ourselves that it could be done.

It took approximately ten years before a themonuclear
explosion on a large scale was attempted on a little
reef called Elugelab, one in the ring of islands which
enclose the lagoon of Eniwetok. During these years,
the war had been ended with a nuclear blast. Work on

fusion had been slowed down, stopped, started a second time and finally advanced to the stage of the test. Difficulties, solutions, proofs of feasibility, proofs that the job cannot be done, new ideas, new difficulties, crowded each other. On the first of November 1952, I could not be on the ship which watched from a distance, the first great thermonuclear explosion. I watched from a greater distance, from California where it was the day of Halloween. I watched a seismograph which was to respond to the shock of the explosion. I knew every argument and every relevant calculation. But I was not certain whether the attempt would succeed. I had convinced others and I had convinced myself. But when I saw the luminous point jump on the photographic film I did not believe that the fleeting motion was real. Only when the photographic strip was developed was I sure that my friends in the Pacific had seen the first truly thermonuclear explosion on earth, half an hour earlier.

It is a pity that even today the rules of secrecy do not permit the open discussion of the development that led to our first success in the massive release of thermonuclear energy. What one writes, depends on the audience. What is written from memory is distorted by hindsight, which is more often foolish than wise. The stories I read about nuclear energy convince me that history, at best, is a patchwork of half-truths.

There is one aspect of that experiment in the Pacific that was exposed to the beneficent light of scientific publication. Glenn Seaborg, with a score of collaborators, collected the radioactive debris from the neighborhood. (The little island Elugelab itself was sunk.) The analysis showed that two new elements had been formed with atomic numbers 99 and 100. In less than a microsecond during the explosion, many neutrons attached themselves successively to some uranium nuclei. Thus were produced the heaviest elements ever yet encountered. By a slower series of succeeding transformations of neutrons into protons, the new elements were made. They were given names: Einsteinium and Fermium. I had nothing to do with the choice of names, but these two great scientists happened to be the mean who gave me the first impetus which eventually led to the explosion of an atoll in the south seas.

In a way, the history of the fusion bomb was not as terrible as that of the fission bomb. The first atomic bomb was hardly finished before it was used on

Hiroshima and Nagasaki. We missed the great opportunity
to finish the Second World War with a mere demonstration.
The idea of atomic weapons was linked in the minds of
people with the ideas of destructive power and of fear.

The hydrogen bomb has not yet been used in anger.
I hope it never will be so used. In fact, it can be
readily adapted to constructive uses. It can cheaply
develop almost unlimited power and it need not generate
excessively dangerous radioactivity. It can be used for
digging harbors or canals. It can break up rock at
great depths so that useful materials can be extracted
in an inexpensive manner. Natural gas and oil and cop-
per can be obtained in this way, and other raw materials
surely will follow. The thermonuclear explosive would
have long since been used for such constructive purposes
except for the formidable psychological barrier which was
raised by Hiroshima.

Another phase of thermonuclear research had quite a
different history of development. As soon as the development of the hydrogen bomb entered its last successful phase the question arose, both among technical people
and administrators, whether one can utilize thermonuclear energy in a slow and controlled manner, rather
than in an explosive arrangement. The difficulties of
making thermonuclear explosions had been greatly over-
estimated. By contrast, the obstacles in connection with
controlled thermonuclear reactions were at first not
recognized and work on the subject started with the
expectation of rapid success. Actually, the problem
of controlled thermonuclear reactions turned out to be
an exceptionally knotty one.

The thermonuclear part is the same whether one
tries to produce an explosion or a slow reaction.
Furthermore, it is not difficult to see that the reaction
can be appropriately slowed down and satisfactorily
governed if only the reacting nuclei, heavy hydrogen
called deuterium, and the even more heavy hydrogen,
called tritium, are introduced at a high temperature
and at an exceedingly low density. What one has to
work with may be designated as a high pressure vacuum.
The word vacuum is used because rather few particles
are present in unit volume. But these particles must
be brought to a temperature of about one hundred million
degrees Fahrenheit, and the violent motion at these
temperatures will produce respectable pressures even
though the number of particles is small.

The first difficulty becomes obvious when one
remembers that at the low densities which are required,
the particles will as a rule collide with the walls of
the confining vessel, rather than with each other.
Even if one succeeds in heating the particles to very
high temperatures, they will lose their energy in the
collision with the walls, which of course, cannot be
kept at a high temperature if they are to remain solid.
One needs, therefore, a container whose walls will not
participate in the characteristics of heat phenomena.
An insubstantial wall of this kind can indeed be made
using magnetic fields for the purpose of confining the
reacting material. The arrangement may be given the
name of "magnetic bottle". Magnetism is in fact the
bottle which contains the thermonuclear fuel.

That such confinement is possible is due to the
fact that at the high temperatures in question, each
hydrogen atom is decomposed into the positively charged
nucleus, and negatively charged electron. Under the
influence of magnetic fields, these charged particles
will not move along straight lines but rather will
spiral along magnetic lines of force. One may arrange
the magnetic lines in a ring-like configuration so that
no open ends are left. In this way the thermonuclear
fuel can be trapped. Alternatively, one may permit open
ends by making an arrangement such that near these ends
the magnetism should have a greater intensity. One
can then show that these more intensive regions at the
two ends of the magnetic bottle reflect many of the
particles which are to participate in the thermonuclear
reaction. The ends of this magnetic bottle are appropriately called magnetic mirrors.

The first discussion of these arrangements for
containing the reactants led in the early 1950's to a
great wave of optimism. This optimism was borne out
when it was found that at least a few charged particles
can be stably and reliably contained by the magnetic
arrangments. Some early experiments even indicated that
by rapidly heating and compressing the material trapped
by magnetic fields, one can produce neutrons which
indeed always occur when thermonuclear reactions take
place.

Then came the first great disappointment.

In the actual experiment, temperatures required for
the fusion reaction were not produced. The neutrons
were due to reactions between fast particles which had

been accelerated by strong local fields. These in turn
arose from instabilities. Energy stored in the gas and
in the magnetic fields was released in irregular macroscopic motions. As a result the magnetic bottle became
leaky and temporarily strong electric fields were produced. This accelerated some deuterons and some nuclear
reactions occurred. But confinement of the hot gas and
steady production of nuclear energy were not attained.

What seemed to be the threshold of success turned
out to be an all but hopeless situation. As long as
little gas was confined, the magnetic bottle held.
But the small amounts are of no use in the production
of energy. As soon as the gas pressure became greater,
magnetic fields started to behave like rubber bands and
the confinement failed.

Instabilities are not uncommon in physics. A pencil
balanced on its tip is an obvious example. Hydrodynamics
is full of instabilities. Irregular flow patterns
develop and give rise, in the end, to turbulent motion.

In the hot ionized gases which we discuss here and
which go by the name of "plasma", instabilities are even
more common. The positive and negative particles can
move almost independently. Furthermore, magnetic
fields are distorted and electric fields are generated.
These may accelerate groups of charged particles having
specific velocities or moving in certain directions.
The patterns of possible instabilities seemed to have
no limit.

The difficulties encountered in plasma physics led
to an important practical result. The research had been
proceeding under raps of secrecy but in 1959, when it
became quite clear that no early success could be expected, the whole area was declassified. In the Second
"Atoms for Peace" Conference in Geneva, a free discussion
took place between all concerned, and ever since that
time the effort has become as international as any other
scientific development.

The interchange of ideas has paid off amply in the
past years. I believe that its long-range effects will
be even more beneficial. We are still far from success
but there is an indication that we are at least nearing
the end of the exploratory phase. Theories concerning
instability have been verified in detailed experiments,
and during the past couple of years no essentially new
instabilities have shown up. Under these conditions,

one may be optimistic, at least to the extent that
in the next few years one may hope to construct a
small model in which at least for a short period, one
can release more electric energy than one had to apply
to begin with in order to get the thermonuclear reaction
going.

To accomplish that we probably will need not only
deuterium, but also the heavier and much more expensive
hydrogen isotope, tritium. The reactions between
deuterium and tritium are much easier to initiate and
the whole process occurs at a lower temperature, thus
reducing all problems.

To get from a first success to an economically
profitable production of controlled nuclear energy, a
long and hard development in engineering will be required. The problems seem to be much more formidable
than in the case of the controlled fission reactions.
But even in the case of fission reactions a quarter of
a century elapsed between the time of Fermi's first
success and the date when fission reactors became
truly competitive on the energy market.

In the meantime, we are already profiting by the
plasma research which is the essential part in the
development of controlled fusion. It is quite clear
that plasmas, that is, hot ionized materials penetrated
by magnetic fields, play a most important role within
stellar bodies. Astrophysics will probably not be
understood until we have completely mastered the behavior of these plasmas. The basic scientific discussion
of astrophysics which gave the first impetus to thermonuclear developments is now the field which is profiting
most from the results of practical thermonuclear research.

Applied science and basic science are indeed
closely interrelated. The fact that work on controlled
thermonuclear reactors is carried out today without
secrecy restrictions has permitted and encouraged this
fruitful interaction.

The explosive applications of thermonuclear energy
are of quite comparable inherent interest. But the results in this field could not be used in a similarly
fruitful manner, due to the fact that this topic is
still kept secret.

Pure science has immeasurably benefited when secret
alchemy was replaced in the late 18th century by open

science.   It is to be hoped that ways can be found in
which open research can be reestablished in all parts
of practical, as well as in all of basic science.
We should use nuclear physics in an open manner for all
practical and also all intellectual purposes.

VLADIMIR K. ZWORYKIN
WITH AN IMAGE ICONOSCOPE

VLADIMIR K. ZWORYKIN

TELEVISION: ITS INDEBTEDNESS AND CONTRIBUTION TO SCIENCE

Vladimir K. Zworykin[*] and Edward G. Ramberg

RCA Research Laboratories, Princeton, New Jersey

CHAPTER I

Television in Myth and History

The desire to extend vision beyond the horizon, to see the unseen, is as old as man's ability to think, to ponder, and to hope. Yet, up to a century ago, the attainment of this objective in any general manner appeared so remote, so improbable, that it was regarded as the province of magic. Efforts to realize vision at a distance on a scientific basis began in the 1870's, even though one of the essential elements, the amplifier, was lacking until De Forest's invention of the audion in 1906.

The advent of the vacuum tube and the vacuum photocell eventually made television a practical possibility. It ushered in the mechanical era of television. In the 1920's numerous mechanical scanning devices of great ingenuity were constructed and demonstrated. Television stations were launched to provide visual entertainment, in emulation of the highly successful sound broadcasting stations. Yet the public appeal of the new medium remained limited; mechanical scanning proved unable to provide the picture quality demanded by the public.

Fortunately, before the mechanical era had ended in disappointment, the ground work for the electronic era had already been laid. Television disappeared from

---

[*]Also at the Center for Theoretical Studies, University of Miami, Coral Gables, Florida.

the market place, but the work in the laboratories proceeded, with changed emphasis. When, at the end of
World War II, the opportunity for a large-scale broadcast service presented itself once again, electronic
television was ready. The subsequent phenomenal growth
of the television industry is a familiar story.

A. The Prescientific Era. In earliest times the
gift of vision unlimited by space or time was believed
to be conferred on favored individuals by the super-
natural powers. Men came from far and wide to consult
and give tribute to the priests of Delphi and to other
oracles. Specific devices, also, were believed to give
the possessor the ability to see beyond the natural
range of his eyes.

The mirror, in particular, is frequently invested
with magic powers in the traditions of different cultures.[1] Some of these traditions, such as the mounting
of mirrors in the lighthouse of Alexandria and on the
Colossus of Rhodes to aid in detecting approaching
vessels at great distances, seem so reasonable that they
may have been founded on fact. Others, telling of con-
cave or convex mirrors which revealed the whole universe,
are clearly fanciful extrapolations derived from the
strange and uncomprehended behavior of actual mirrors.
The frequency with which this notion recurs in the tales
of different peoples reflects the universal yearning to
exceed the limits of natural vision and to see what goes
on beyond the horizon.

Even long after the belief in magic had ceased
among the educated we find references in fiction to remote vision by magic powers. An interesting example is
Sir Walter Scott's My Aunt Margaret's Mirror, published
as one of the Waverley novels around 1825. Scott lets
his heroines, who are attempting to locate an unfaithful
husband, consult a physician with the reputation of a
conjurer. The conjurer leads them into a chamber containing a very tall and broad mirror, where, in Scott's
words:

> Suddenly the surface assumed a new and
> singular appearance. It no longer simply
> reflected the objects placed before it, but
> as if it had self-contained scenery of its
> own, objects began to appear within it, at
> first in a disorderly, indistinct, and miscel-
> laneous manner, like form arranging itself
> out of chaos; at length, in distinct and

defined shape and symmetry. It was thus
that, after some shifting of light and darkness over the face of the wonderful glass,
a long perspective of arches and columns
began to arrange itself on its side, and a
vaulted roof on the upper part of it; till,
after many oscillations, the whole vision
gained a fixed and stationary appearance,
representing the interior of a foreign
church.....

Scott's description is here strikingly reminiscent
of what a television viewer might observe during the
warming-up time of a receiver.

While reference to the employment of magic for the
extention of human sight is not uncommon in literature,
relatively few writers have given television a share in
their view of the future. An exception is George
Bernard Shaw. In the third act of Back to Methuselah,
which is supposed to take place in A.D. 2170, Shaw lets
the President of the British Islands communicate with
his associates in various parts of the world by two-way
videophone. For this purpose, in the presidential office
"the end wall is a silvery screen nearly as large as a
pair of folding doors." To establish visual connection
the President and his counterpart are provided with a
switchboard and dial on which they set the desired
number.

Back to Methusaleh was published in 1921. By this
time television had become a distinct technical possibility, although several years were to pass before the
first crude demonstrations. Nevertheless, what may well
have been the first cartoon joke at the expense of television had appeared in Punch some forty years earlier,
so that the idea of seeing at a distance by electrical
means must have been familiar to a considerable fraction
of the population. Clearly, the history of the ideas on
which modern television is based goes back far beyond
the publication date of Shaw's play.

B. The Era of Speculation. As we turn to the
development of television on a scientific basis we encounter a period lasting approximately forty years which
we may properly call the era of speculation.[2]* In this

---

*Hogan (reference 2) applies the term "speculative
period" to the interval ending in 1890. We extend it

time numerous useful and ingenious proposals were advanced, even though essential factors for their practical realization were absent.

The first system for the electrical transmission of pictures known to the authors was described by Alexander Bain in 1843.3** It was designed specifically for the transmission of type face. In Bain's system signals were transmitted along parallel wire lines connected to the insulated teeth of a wire comb. The comb was translated across a plate containing short, parallel wire sections in an insulating matrix while the type face connected to a battery was pressed against the opposite side of the plate. At the receiver a similar comb was translated across chemically treated paper, which, like the other pole of the battery, was connected to ground. Thus, whenever a tooth of the comb at the transmitter passed over a section of type face, current passed through the corresponding tooth of the comb at the receiver and effected discoloration of the paper.

Bain's system was never built. However, Frederick Collier Bakewell, about four years later, transmitted written matter over a single telegraph line; the original was written with insulating ink on metal foil placed on the mantel surface of a drum rotated by clockwork. The foil was explored by a probe translated slowly parallel to the axis of the drum, so that the written matter was scanned along a series of closely spaced, parallel lines.

Bakewell's apparatus resembles in many respects modern facsimile -- the electrical transmission of printed material, photographs, or drawings. It also exhibits certain features, such as scanning, which have been of primary importance to television. However, neither Bain nor Bakewell appears to have had the remotest idea of applying their techniques to the instantaneous transmission of pictures of action. This was natural, since they lacked not only the means of

----

here up to approximately 1920, when workers had the essential elements for the practical realization of television at their disposal.

** See Korn and Glatzel (reference 3). The historical references to the early history of television, where not otherwise indicated, are derived from this source, Hogan (reference 2), Goebel (reference 7), and Jensen (reference (18).

transmitting pictures at sufficient speed to permit the
subjective merging of successive frames into a continuous
representation, but also the essential photoelectric
means for converting variations in light intensity into
variations of current or voltage. Edmond Becquerel's
discovery of the photovoltaic effect a few years earlier
(1839) appears to have excited little notice among in-
ventors.

The effect of the observation of the photoconduct-
ivity of selenium by May and Willoughby Smith in 1873,
on the other hand, was pronounced. It is true that the
first description of a television system, by G. R. Carey
of Boston in 1875, did not contemplate use of this
material. Carey proposed that a picture of the object
to be transmitted be projected on a photosensitive insulating surface, such as a layer of silver halide.
Terminal pairs, in the form of closely spaced platinum
wires of a multiplicity of electrical circuits, were
buried in the photosensitive layer so that they would
be short-circuited with the partial reduction of the
silver halide to silver by the incident light. Each
circuit controlled, eventually through a relay, an
electric light source in a panel of sources arranged in
the same geometrical order as the terminal pairs in the
photosensitive layer. Thus a projection of a light pattern
on the photosensitive plate would result in the formation
of a similar light pattern on the "receiver" panel.

Clearly, Carey's system as described above was not
operative. However, it corresponds in essence to a
method employed even at the present time to animate advertising signs. Furthermore, it was the only early
system that could have functioned with photosensitive
materials having a response as slow as the early selenium
cells. In Carey's system, with a separate channel for
each "picture element," the cell response had to be
only fast enough to permit the observation of continuity
of motion. In the nonstorage scanning systems, subsequently introduced, the rise time and decay of the
response had to be faster by a factor equal to the number
of picture elements in order to prevent the washing out
of the picture by overlap of the picture signals from
one picture element to the next.

A selenium cell panel serving the same purpose as
Carey's photosensitive plate was built by Ayrton and
Perry in England in 1877. In 1906 Rignoux and Fournier
built an array of 64 selenium cells, each coupled to a
shutter in front of an extended light source, and used

this system to transmit simple patterns. However, the main line of development of television lay in another direction -- at least with respect to the method of signal transmission. Even as early as 1880 it was realized that at least 10,000 picture elements would be needed to reproduce pictures with adequate detail.[4] The employment of parallel transmission lines and shutter or relay devices in such numbers appeared clearly impractical.

The remedy lay in the application of the scanning principle, applied at a much earlier time by Bakewell to facsimile transmission. The first suggestion of its application to television appears to have been made in 1878 by the Portuguese physicist de Paiva, who suggested that the picture to be transmitted be projected on a selenium-coated metal plate scanned by a metal point. The metal point and plate were connected in series with a battery and relay. The relay controlled the current through an electric light bulb which was translated in the same manner as the metal point. Since the bulb would light up whenever it occupied positions corres-ponding to the bright portions of the picture, it would trace out a replica of the latter.[5]

This and innumerable later proposals existed, of course, only on paper. Thus a contemporary of de Paiva, M. Senlecq, first suggested the scanning of a picture formed on a ground-glass screen with a small selenium cell and, at a slightly later date, the scanning of an array of small, stationary selenium cells with a metal brush. In either case the picture was to be reproduced on paper, either by a soft pencil pressed against it by a magnetic armature or by electrochemical action. Thus Senlecq was actually dealing with facsimile rather than with television. Few of the early authors were aware of the essential distinction between television and facsimile.

However, a number of devices and principles, which proved important to the later development of television, were developed in this period. Paul Nipkow's invention of the scanning disk in 1884 not only gave television its first practical means for transmitting pictures at sufficient speed for continuous observation over a single channel, but also presented it with the instrument which had the greatest survival value of all mechanical scan-ning devices.

The Nipkow disk (Fig. 1) is a round disk with a

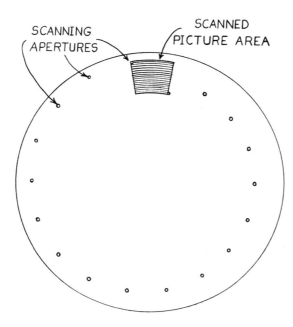

Fig. 1.   The Nipkow disk.

series of apertures placed along a spiral. If a picture corresponding in width to the angular separation
of successive apertures and in height to the maximum
radial separation of the apertures is projected on the
rotating disk, the apertures scan the picture in a
succession of parallel lines. Thus, if the light transmitted by the apertures falls on a photocell, the photocurrent generated in the cell will correspond to the
brightness variation of the picture along the scanning
lines. If a second Nipkow disk, whose rotation is
synchronized with the first, is placed at the receiver
in front of a light source modulated by the current from
the photocell, the original picture will appear reproduced in the plane of the disk; the rotating disk
apertures will light up with a brightness proportional
to that of any particular element of the picture at the
moment when they occupy a position corresponding to that
particular element (Fig. 2).

Nipkow and his contemporaries suggested many other

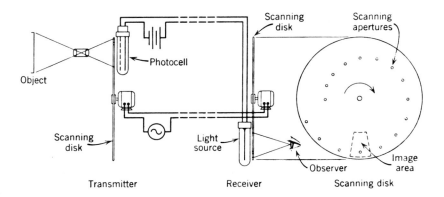

Fig. 2. Mechanical television system employing Nipkow
disk in pickup and receiver. (From V. K. Zworykin and
E. G. Ramberg, Photoelectricity, Wiley, New York, 1949.)

techniques which were investigated experimentally at a
later period. Nipkow himself proposed the employment
of lenses in the scanning-disk apertures to increase
the amplitude of the picture signal and the use of the
Faraday effect for inertia-free light modulation in the
receiver. LeBlanc, in 1880, suggested the employment
of two mirrors oscillating about mutually perpendicular
axes for scanning the picture. Weiller, in 1889, developed the mirror drum, with as many mirrors as there
were scanning lines, each tilted by a slightly greater
angle than the preceding mirror with respect to the
axis of rotation. Thus the rotation produced the
horizontal deflection of the picture element, and the
increasing tilt, the vertical deflection. Many other
mechanical methods of scanning, as well as practical
means for synchronizing scanning in the transmitter
and receiver (e.g., by the use of phonic wheels, as
suggested by Nipkow), belong to the contributions of
this period.

However, the inventions of the speculative period
were not limited to mechanical devices. The cathode-
ray oscilloscope, developed by F. Braun in 1897 for the
study of the time variation of electric currents, was
utilized for picture reproduction by M. Dieckmann and
G. Glage in 1906. The objects in their apparatus were
metal patterns which were explored by twenty contact
brushes replacing the scanning apertures on a Nipkow
disk. The Nipkow disk was coupled mechanically to a

generator delivering horizontal sawtooth current to the
horizontal deflection coils on the cathode-ray tube and
to the contact brush on a slide-wire potentiometer
delivering the vertical deflection currents to the
vertical deflection coils. Whenever the brushes on the
Nipkow disk made contact with a point of the metal
pattern, they caused current to flow through an electro-
magnet which deflected the electron beam in the oscilloscope so as to miss an aperture in its path. Thus conducting portions of the pattern were reproduced dark on
a luminous background on the oscilloscope screen. Since
a complete scan (a rotation of the Nipkow disk) was
completed in 0.1 second, displacements and rotations of
the pattern could be followed readily on the screen.

The transmitter of Dieckmann's apparatus was such
that, viewed as a whole, it was a facsimile system rather
than a television system. Since there were no means for
amplifying photocurrents at the time, the facsimile-
type transmitter employing contact brushes alone enabled
Dieckmann to realize his system physically; the apparatus
employed in 1906 to transmit written matter and simple
drawings is at present in the German Technical Museum
in Munich. Although the system as a whole was a facsimile system, the receiver (Fig. 3) was an operative
electronic picture reproducer.

Shortly after the publication of the patent granted
to Dieckmann and Glage, a patent was issued, in 1907,
to B. Rosing of St. Petersburg for a television system
employing a cathode-ray tube receiver. It employed at
the transmitter two mirror prisms rotating about mutually
perpendicular axes, which effected the scanning displacement of the picture to be transmitted across an aperture.
This feature was not novel, of course. However, Rosing's
system possessed the important innovation of replacing
the very slow selenium photoconductive cells employed
as light-sensitive elements (placed behind the aperture)
by inertia-free alkali phototubes. Thus one of the
basic obstacles to television over a single signal
channel was overcome. The other difficulty, namely the
fact that the photocurrents were much too weak to
actuate any receiving apparatus without amplifying means,
prevented the practical operation of Rosing's apparatus,
just as in all the single-channel television systems of
his predecessors. Nevertheless, the apparatus actually
was constructed and exhibited in St. Petersburg in 1910.
In Rosing's system both the horizontal and the vertical
deflection of the beam were effected by the current

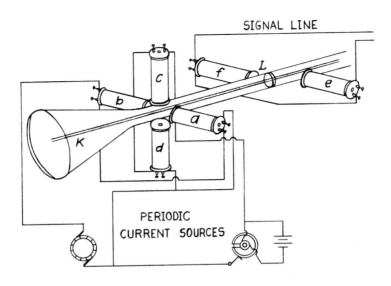

Fig. 3.   Cathode-ray tube receiver of M. Dieckmann and G. Glage.   K, electron beam; a, b, vertical deflection coils; c, d, horizontal deflection coils; e, f, intensity modulating deflection coils; L, diaphragm limiting transmitted beam current.   (M. Dieckmann and G. Glage, German Patent 190102, Sept. 12, 1906.)

induced in pickup coils, series-connected to the deflection coils of the cathode-ray tube, by permanent magnets associated with the individual mirrors of the mirror drums.   Otherwise the picture-reproducing system was essentially the same as that of Dieckmann and Glage.

The first suggestion of eliminating mechanical devices altogether from a television system was made by A. A. Campbell-Swinton in 1908.   Campbell-Swinton saw in the deflection of electron beams in the transmitter and the receiver (by electromagnets actuated by alternating currents of two greatly differing frequencies) the only method of achieving adequately high picture definition.   He gave his ideas more precise form in a lecture to the Roentgen Society in London in 1911[6] (Fig. 4).   The novel part of his system, as compared to those of Dieckmann and Rosing, is, of course, the transmitter, or, in modern parlance, the camera tube.   It employed a mosaic of mutually insulated photo-

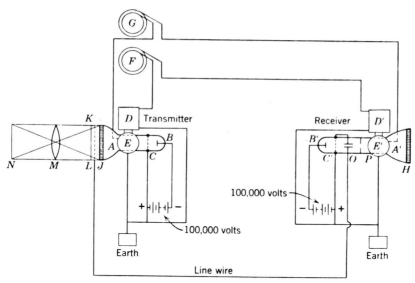

Fig. 4.   Electronic television system proposed by
Campbell-Swinton. (From V. K. Zworykin and G. A. Morton,
Television, 2nd Ed., Wiley, New York, 1954, Fig. 7.13.)

sensitive elements (e.g. cubes of rubidium), which was
scanned on one side by an electron beam. The other side
of the mosaic formed a wall of a receptacle "filled with
a gas or vapour, such as sodium vapour, which conducts
negative electricity more readily under the influence
of light than in the dark." The same receptacle contained a collector screen for this negative charge. A
line wire connected this collector screen to the control
element in the receiver, so that the magnitude of the
current received by the collector screen determined the
brightness of the scanning spot on the receiver screen.

In greater detail, the manner in which Campbell-
Swinton pictured the operation of his camera tube is
indicated by the following sentence:

In the case of cubes (of rubidium) on
which no light is projected, nothing further
will happen, the charge dissipating itself in
the tube; but in the case of such of those
cubes as are brightly illuminated by the projected image, the negative charge imparted to
them by the cathode rays will pass away
through the ionized gas along the line of

the illuminating beam of light until it
reaches the screen L, whence the charge
will travel by means of the line wire to
the plate O of the receiver....

He regarded as one of the chief advantages of his
system the fact that he employed a multiplicity of photoelectric cells in parallel, each of which was called upon
to deliver an electric pulse once in a picture period
(e.g. 1/10 second). This was unlike:

...other arrangements that have been suggested,
in which a single photoelectric cell is called
upon to produce the many thousands of separate impulses that are required to be transmitted through the line wire per second,
a condition which no known form of photoelectric cell will admit.

The parallel arrangement of photosensitive elements,
or the photosensitive mosaic, proved indeed an essential
feature of high-sensitivity television camera tubes,
though for other reasons than those given above. With
the introduction by Rosing of the high-vacuum phototube
into the television art, the slow speed of response of
photoelectric devices ceased to be a barrier. This had
taken place even before the writing of Campbell-
Swinton's article.

Needless to say, Campbell-Swinton's television
system was never constructed. Furthermore, familiarity
with his proposal would scarcely have altered the ver-
dict reached by Arthur Korn, a pioneer in the field of
facsimile, after an exhaustive review of earlier television efforts in 1911: "With means now known to us,
practical television, employing a single line or a small
number of lines, cannot be realized."

At the time this was unquestionably a reasonable
position. Nevertheless, the mental effort and experimentation applied to the television problem in the period
of speculation contributed materially to the speed with
which it was solved when the means essential to the
realization of television became available.

It is also of some interest to note that the only
kind of television with which the speculative period
concerned itself was point-to-point television; the
whole idea of broadcasting is of more recent origin.
We cannot do better than to terminate this account of

the speculative period with the vision of one of its
outstanding exponents, B. Rosing, as expressed in an
article in the French journal <u>Excelsior</u> published a-
round 1910.[7]*

> The range of application of the tele-
> phone does not extend beyond human conver-
> sation. Electrical telescopy will permit
> man not only to commune with other human
> beings, but also with nature itself. With
> the "electric eye" we will be able to pene-
> trate where no human being has seen. The
> "electric eye," fitted with a powerful lamp
> and submerged in the depths of the sea, will
> permit us to read the secrets of the submarine
> domain. If we recall that water covers
> three quarters of the earth's surface, we
> readily realize the infinite extent of man's
> future conquests in this portion of his
> domain, till now inacessible to him. From
> now on and in all future times we can imagine
> thousands of electric eyes traveling over the
> floor of the sea, seeking out scientific and
> material treasures; others will carry out
> their explorations below the earth's surface,
> in the depths of craters, in mountain cre-
> vasses, and in mine shafts. The electric eye
> will be man's friend, his watchful companion
> which will suffer from neither heat nor cold,
> which will have its place on lighthouses and
> at guard posts, which will beam high above
> the rigging of ships, close to the sky. The
> electric eye, a help to man in peace, will
> accompany this soldier and facilitate commu-
> nication between all members of human society.
> Yet we need not fear that it might pry into
> the private life of families, as some might
> think; it is incapable of seeing through
> walls.

C. <u>The Mechanical Era</u>. The speculative era came
to an end when all the means essential to translating
television into practice became available to workers
in the field. These were, in particular, the almost
inertia-free photoemissive effect, utilized in the
vacuum phototubes of Elster and Geitel and others; the
electronic amplification of voltages and currents,

---

*Quoted in Korn and Glatzel (reference 3).

based on the three-element vacuum tube introduced by
De Forest in 1906; and light sources permitting rapid
modulation, such as the negative-glow neon lamp of
D. MacFarlan Moore, introduced in 1917. By 1920 these
factors had been combined to make television workable.

It may occasion surprise that the cathode-ray tube,
whose ability to reproduce images was demonstrated by
Dieckmann as early as 1906, was not generally considered
an important factor in television in 1920. This may be
attributed to the fact that the early cathode-ray tubes
had to rely on gas focusing of the scanning beam; hence
they were short-lived and limited in spot sharpness and
permissible speed of deflection. Long-lived tubes
delivering sharp, bright images at high scanning speeds
became possible only with the application of electron-
optical principles to the focusing of electron beams in
vacuum.

Accordingly, mechanical methods of scanning, both
in the transmitter and in the receiver, were generally
preferred in the period beginning about 1920. They
dominated the field up into the early thirties, when
they were replaced by electronic scanning techniques,
first in the receiver and then in the transmitter.
Thus it appears appropriate to designate the relatively
brief period from 1920 up into the 1930's as the
mechanical period of television.

The year 1920 also marks the beginning of commercial sound broadcasting in the United States. The
enormous success of radio broadcasting for entertainment
purposes redefined the objectives of television. The
prospect of television broadcasting eclipsed all other
uses. Only one other objective was pursued to any
extent during the mechanical era: two-way video
communcation. The early studies by H. E. Ives and his
associates at the Bell Telephone Laboratories, carried
out in particular in the years 1927-1930, were largely
concerned with television as an extension of telephone
communication. In Germany G. Krawinkel demonstrated a
two-way television system at the radio exposition in
Berlin in 1929, and from 1936 to 1940 the German Post
Office provided public, two-way videophone service between Berlin, Leipzig, Nürnberg, and eventually,
Munich.[8] With these minor exceptions attention was
concentrated on broadcast television for entertainment
purposes.

In some respects, the mechanical era may appear as

an unproductive diversion from the main path of development. The mechanical scanning techniques, utilizing
scanning disks, rotating mirrors and prisms, mirror
screws, rotary commutators, etc., though developed to a
high level of perfection, had limited permanent significance in television. On the other hand, the same period
was extremely productive in formulating the requirements
of television communication and developing the necessary
circuitry.

To begin with, the number of lines in the picture,
which determined the discrimination of detail in a
vertical direction, was limited by the sensitivity of
the transmitting equipment and by problems of mechanical
construction. Thus in 1926 J. L. Baird, in England,
showed 30-line pictures scanned at a frequency of five
pictures per second. In the following year H. E. Ives
and his co-workers at the Bell Telephone Laboratories
transmitted 50-line pictures between Washington and New
York at a rate of 17.7 pictures per second. The first
television standards established by the German Post
Office, in 1929, prescribed 30 horizontal lines, an
aspect ratio (width:height) of 4 : 3, and a picture frequency of 12.5 per second. However, technical improvements forced an early upward revision in the line number
and frame frequency. By 1932 the Radio Corporation of
America transmitted pictures from New York with 120
scanning lines and a frame repetition frequency of 24
per second.

It became increasingly clear that technical factors
alone would not indefinitely prescribe reasonable upper
limits to the frame frequency and line number of television pictures. Hence tests were carried out with
pictures artificially broken up into picture elements
of uniform intensity and varying size and with others
projected intermittently at varying frequency and with
varying duty cycle.[9] The structured pictures were viewed
by a number of observers, who determined the closest
distance from the picture for which the detail appeared
satisfactory. It was found that this corresponded
approximately to the distance for which a line spacing
subtended the minimum angle of resolution of the average
eye, or about two minutes of arc. This finding, together
with the observation that the closest viewing distance
for which the entire picture could be seen clearly and
simultaneously was four times the picture height, led
to the standards of 400 to 600 lines prescribed in most
countries of the world today. The aspect ratio of 4 : 3

was taken over from motion-picture practice.

The tests with intermittently projected pictures
sought to determine the lowest field frequency for which
the viewer perceives no disturbing flicker effects.
This increases with picture brightness and is considerably higher than the frequency required to give an
appearance of continuity of motion. Thus in motion-
picture practice it is customary to project each picture
twice, making the field frequency (48 per second) twice
as large as the frame frequency (24 per second). Since,
in television, the individual fields are not presented
to the eye as a unit, but are built up by a process of
line scanning, the flicker tests on projected pictures
were eventually supplemented by tests with television
pictures of varying brightness and field frequency
formed by cathode-ray tubes. These tests showed that the
field frequency of 48 per second adopted in motion-
picture practice gave an entirely flicker-free presentation for picture brightnesses up to 20 foot-lamberts.
To minimize the disturbing effects of hum (cross-talk
from the power lines into the deflection circuits),
the field frequencies were generally set equal to the
power-line frequencies in the service area, or 50 per
second in Europe and 60 per second in America. The
relatively high field frequency adopted in America has
proved particularly fortunate, since it provided flicker-
free viewing even at the very high picture brightnesses
realized in more recent years.

As in motion picture practice, the frame frequency
was chosen to be half as large as the field frequency.
This was made possible by interlaced scanning. In
alternate fields all the odd lines and then all the even
lines of a complete frame are scanned. Interlaced
scanning has the advantage that the fineness of picture
detail in a vertical direction and the freedom from
flicker are determined by the total line number in the
scanning pattern and the field frequency, respectively,
whereas the scanning speed and consequently the frequency band required to transmit the picture information
are determined by the product of the line number and the
smaller frame frequency.

The above-mentioned studies, which formed a sound
basis for the standardization of the number of the scanning lines and the field frequency of the television
picture, were supplemented by careful theoretical studies

of picture reconstruction in television.[10]* These
established, among other things, the relationship between horizontal resolution and the frequency bandwidth
of the transmission channel between picture pickup and
reproducer and the influence of the size of the scanning
spot in either on the reproduced picture.

Finally, satisfactory methods for generating and
utilizing synchronizing signals were worked out. These
assured the proper framing of the received pictures.
Since the synchronization methods were intimately linked
with the mechanical scanning methods employed, they have
not carried over into current practice. However, they
indicate a clear understanding of television synchronization requirements.

Although the mechanical era did not succeed in
creating satisfactory television systems, it did provide
a clear formulation of the problems of television.
Furthermore, to a very considerable extent, it saw the
development of the auxiliary means required for any
successful television system, mechanical or electronic.

D. The Electronic Era. The mechanical era of
television merged into the electronic era as the limitations of mechanical scanning systems came to be
recognized and electronic systems were made available
to replace them. In practical form, this occurred first
for the receiver (about 1930) and somewhat later, in
the middle thirties, for the transmitter. Several individual workers had, of course, committed themselves
to an entirely electronic solution of the television
problem at a much earlier date.

We have already seen that a form of cathode-ray
tube had been employed for picture reproduction by
Dieckmann and by Rosing in the speculative era of
television. In these early tubes the scanning spot
intensity was modulated by deflecting the electron beam
across an aperture. No method of beam focusing other
than gas focusing was known or employed. The modern
practice of beam intensity modulation by an axially
symmetric grid was adopted in experiments in electronic
television demonstrated by the senior author in 1924
and described in a slightly earlier patent appli-

---

*See Mertz and Gray (reference 9); for a more compre-
hensive review of the subject, see Zworykin and Morton,
Chapter 5 (reference 10).

cation.[11][*] Five years later he described an essentially modern television viewing tube which incorporated grid modulation, a hard vacuum, and (electrostatic) electron-optical focusing.[12][**] He named this tube a "kinescope" (Fig. 5). A fuller description of the operation of improved tubes of this type, in terms of the language of electron optics, was given in a few years later;[13][†] a general theoretical basis for this new science had been created in the meantime by Hans Busch.[14][††]

The cathode-ray viewing tube speedily replaced mechanical picture reproducers in the early 1930's. The Farnsworth "oscillite"[15][§] and the picture tubes built by M. von Ardenne[16][§§] in Germany incorporated improvements similar to those realized in the Zworykin kinescope. Magnetic deflection, effected by electronically generated sawtooth currents, was employed quite generally. Correct timing was achieved by the application of synchronizing pulses to the control oscillator of the deflection generator, as at present.

The problem of developing a satisfactory electronic pickup device proved more difficult. Although Campbell-Swinton had pointed out that the realization of a workable television system demanded the adoption of electronic scanning in both the transmitter and the receiver, there is no evidence that he ever constructed such a system. However, late in 1923 the senior author, then associated with the Westinghouse Electric Corporation, succeeded in transmitting a crossmark with a television system which utilized the tube shown in Fig. 6 as transmitter and a cathode-ray tube as receiver.[17][*]

The tube in Fig. 6 employed a thin aluminum signal plate which was oxidized on one side. The insulating

---

[*] V. K. Zworykin, U. S. Patent 2,141,059, Dec. 20, 1938 (filed Dec. 29, 1923).

[**] See Zworykin (reference 11).

[†] See Zworykin (reference 12).

[††] See Busch (reference 13).

[§] See Dinsdale (reference 14).

[§§] See von Ardenne (reference 15).

Fig. 5.   An early kinescope.   (Zworykin, 1929).

(oxidized) side was photosensitized and faced a metal
grill which served as collector. The picture to be
transmitted was projected through the grill on the photosensitive layer and built up on it, by photoemission, a
charge image. An electron beam issuing from the gun at
the narrow end of the tube scanned the signal plate and
penetrated the oxide layer so as to establish a conducting path between the signal plate and the photosensitive elements on the opposite side of the layer.

The tube just described was capable of transmitting
only very crude patterns. On the other hand, it demonstrated not merely the possibility of transmitting pictures by electronic means, but illustrated also the
important principle of storage: the charge which gave
rise to the picture signal was not just that liberated
by photoemission from the picture element in question
at the instant of scanning, but was built up on the
insulated photosensitive "mosaic" during the entire
period which had elapsed since the preceding scan.
Ideally, the storage principle permits an increase in
sensitivity over a nonstorage device by a factor equal
to the number of picture elements in the transmitted
scene, or, roughly, the square of the number of scanning
lines. For modern television pictures this factor lies
between 100,000 or 1,000,000. The transmission of
television pictures under a wide range of illumination,
indoors and outdoors, has become possible only through

Fig. 6. Early Storage type television camera tube.
(Zworykin, 1923).

use of the storage principle.

The iconoscope,[18]* the first storage-type television
tube to find practical use, is shown schematically in
Fig. 7. In it, the picture to be transmitted is projected on a mosaic consisting of silver globules photosensitized with cesium and deposited on a mica sheet.
The back of the mica is metalized to form the signal
plate. In contrast to the tube in Fig. 6 the beam now
scans the insulated side of the photosensitive target
and returns its surface to an equilibrium potential
by secondary emission, both photoelectrons and secondary
electrons being collected by a metal coating on the
tube envelope. Since the charge deposited by the beam
at any point of the mosaic depends on the charge built
up at that point by photoemission during the preceding
frame period, the current capacitatively induced in the
signal plate lead corresponds to the intensity of
illumination of the scanned element of the mosaic.

Iconoscopes of the above type were constructed by
the Radio Corporation of America in 1931 and after 1934
rapidly displaced mechanical picture-pickup devices.**
Later developments, culminating in the image orthicon,
aimed at a more complete utilization of the gain in
sensitivity permitted by the storage principle and at the

---

*This method of producing a great number of electronic
elements (photocell with individual condenser), could
be considered as a forerunner of present day integrated
circuit technology.

**See Zworykin (reference 16).

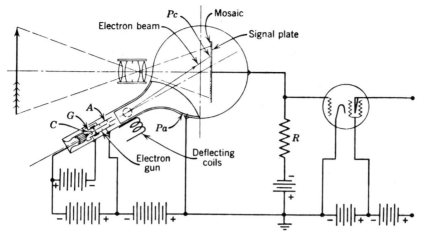

Fig. 7. Diagram of single-sided iconoscope (Zworykin,
reference 16). C, cathode; G, grid; A, anode; Pa,
photocell anode; Pc, photocell cathode; R, signal
resistance.

elimination of residual drawbacks of the iconoscope.

      Besides the iconoscope two nonstorage electronic
pickup devices were developed at an early stage. One
of these is the cathode-ray tube flying-spot scanner.
Suggested in a Zworykin patent application filed in
1923,[19]* flying-spot scanners of this type were built
by M. von Ardenne, both for live transmissions and for
film pickup.** In the flying-spot technique a scanning
pattern of uniform intensity, described on the face of a
cathode-ray tube, is imaged onto the subject to be transmitted, and the light passing through (for film) or
reflected by the subject (for a live subject) is collected by a phototube. The photocurrent, which at any
instant is proportional to the transmission or reflect-
ance of the point being scanned, generates the picture
signal. If the light emission of the phosphor persists
longer than the transit time of a single picture element,
the scanning spot is stretched into a line, and the
horizontal resolution is reduced correspondingly. Early
cathode-ray tube flying-spot scanners suffered from this
defect as well as from the low sensitivity of the available phototubes. For these reasons they did not come

_____

*U. S. Patent 2,141,059.

**See von Ardenne (reference 15).

into practical use until very much later, when efficient short-persistence phosphors and highly sensitive multiplier phototubes could be utilized.

The other nonstorage electronic pickup device which has been applied quite widely both in broadcasting and in industrial television is the image dissector. Here the image of the scene to be transmitted is projected on a photocathode, and the stream of photoelectrons issuing from the cathode is deflected by crossed magnetic fields over an aperture in such fashion that the aperture selects, in succession, electrons from different picture elements along a scanning line. The electrons passing through the aperture are collected and generate the picture signal. This technique, first described and demonstrated by Dieckmann and Hell,[20] was perfected by Philo T. Farnsworth.[21] Farnsworth's important addition was a longitudinal magnetic focusing field which formed a sharp electronic image of the scene in the plane of the aperture. Without this refinement only crude images could be transmitted. Farnsworth's image dissector was introduced in 1934. Later types were greatly improved by the replacement of the collector by an electron multiplier. In this manner they achieved the maximum sensitivity of which a nonstorage television pickup device, operated with available photocathodes, is capable.

The entire technical development which has been described took place with the realization and perfection of a broadcast television service in view. Temporarily, in World War II, this motive was supplemented by the potential uses of television techniques in warfare. In the accomplishment of its main objective television proved remarkably successful. Thus in the United States a majority of households enjoyed television programs originating at the cultural centers of the country within seven years after the establishment of television broadcasting services at the end of World War II. A comparable spread of the new medium of entertainment and instruction, eventually delayed by unfavorable economic circumstances, has taken place for most of the rest of the world. Everywhere television is becoming a part of the prevailing cultural pattern.

CHAPTER II

The Physical Basis of Electronic Television

All technological developments rest, ultimately,
on physical science. In television the two basic
problems to be solved are the conversion of a visual
image into an electrical signal at the transmitter,
and the reconversion of the signal into a visual image
in the receiver. We shall pay less attention to the
problem of conveying the signal from the transmitter to
the receiver, since it is encountered in many different
forms of long-distance communication.

A. <u>Photoelectricity</u>. The generation of an electrical signal in response to a change in light requires
the use of a photoelectric effect. The two photoelectric effects of importance to television are photoconduction and photoemission. We have already seen that
the discovery of photoconduction in grey crystalline
selenium by Willoughby Smith and his assistant May
fired the imagination of early television pioneers, but
that the slow response of this material deprived it of
usefulness in this area. Photoconduction -- the increase
in the electrical conductivity of materials in response
to illumination -- became important to television at a
much later date, with the development of the vidicon
camera tube in 1949.[22]

The photoemissive effect, on the other hand, was of
crucial importance for the practical realization of
television. It was discovered by Heinrich Hertz in
1887, as a by-product of his classical researches on
electric waves and oscillations.[23] He noted that the
spark produced across a gap in a secondary circuit,
employed as measure of electric wave amplitude, could
be lengthened when the gap terminals were exposed to
the spark produced by the primary circuit. A series of
tests established beyond doubt that the ultraviolet
component of the light from the primary spark was
responsible for the effect and that this was greatest
when the ultraviolet light fell on the negative terminal
of the gap.

Painstaking research by Hertz's successors demonstrated that the photoemissive effect consisted of the
emission of negatively charged particles by the illuminated electrode. Noting that the effect was most pronounced for chemically electropositive metals, such as

aluminum, zinc, and magnesium, Julius Elster and Hans
Geitel proceeded to test the still more electropositive
alkali metals, sodium and potassium, amalgamating them
with mercury to retard their oxidation.[24] It was found
that these metals were sensitive not only to the ultraviolet, but also to visible light. Elster and Geitel
took the important further step of enclosing the cathode
and anode (the positively biased collector of the photoemission) in a vacuum envelope, permitting even highly
photosensitivity indefinitely. By 1912 these two
scientists had succeeded in developing, in the potassium
hydride phototube, the first practical phototubes for
visible light, with sensitivities some hundred times as
great as for tubes with pure potassium photocathodes.[25]

     In the meantime, the understanding of the photo-
emissive effect had made great advances. Lenard,[26]
separating out by a small aperture a thin pencil of the
photoemission from an aluminum photocathode enclosed in
a vacuum vessel demonstrated that it was deflected by a
magnetic field to the same extent as the negative emission from any metal cathode in a gas discharge tube.
J. J. Thomson,[27] measuring the deflection of the latter
emission by electric and magnetic fields, had shown it
to consist of identical negatively charged particles
with a mass less than a thousandth of that of the lightest atom. More precise measurements of the charge e
and mass m of these particles, which came to be called
electrons, gave the values

$$e = 1.60 \cdot 10^{-19} \text{ coulomb, } m = 9.11 \cdot 10^{-31} \text{kilogram}$$

It could thus be concluded that photoemission, also,
consisted of electrons.

     Further studies by Lenard and others established
empirically that
　　1.　The number of electrons released per unit time
　　　　at a photosensitive surface is directly pro-
　　　　portional to the intensity of the incident
　　　　light; and
　　2.　The maximum energy of the electrons
　　　　released from the surface is independent of
　　　　the intensity of the incident light, but in-
　　　　creases linearly with the frequency of the
　　　　light.
The second, rather surprising, rule received an interpretation from Albert Einstein[28] in 1905. Einstein made
use of Planck's postulate[29] that light was emitted in

quanta, or energy packets, hν, where ν is the light
frequency and h is a universal constant with the value

$$h\nu = 6.62 \cdot 10^{-34} \text{ joule-second} \quad .$$

He suggested that, in the interaction of light with a
photoemissive material, quanta of light transferred
their entire energy to individual electrons of the
material. If the resultant energy of the electron
sufficed to overcome the forces which bound it to the
material, the electron could be released as a photoelectron with a maximum kinetic energy

$$\frac{mv^2}{2} = h\nu - W \quad .$$

Here W is the energy which the electron must expend to
free itself from the body of the emitting substance --
a material constant known as the work function of the
substance.

It is clear that, for any photoemission to take
place, the quantum energy of the exciting radiation,
hν, must exceed the work function W; for efficient
photoemission, hν must be larger than W. Within the
visual spectrum, the value of the light quantum hν
ranges from 1.6 electron volt in the extreme red to 3.2
electron volt in the blue-violet, whereas the work
functions for pure metals range from 6.3 electron volt
for platinum to 1.8 electron volt for the alkali metal
cesium. Clearly, photoemissive surfaces exhibiting
high sensitivity throughout the visual spectrum had to
be complex in nature.

The immediate goal of research in photoemission became thus the lowering of the work function and the
simultaneous increase of the quantum efficiency -- the
fraction of incident light quanta which result in the
emission of a photoelectron -- within the visual spectrum. Elster and Geitel's potassium hydride cell,
with a maximum quantum efficiency of about 1 percent at
the blue end of the spectrum and a long-wave threshold
in the yellow (corresponding to 2.1 electron volt), was
only a beginning. Some of the important landmarks in
the subsequent development are the discovery of the
silver-oxy-cesium photocathode, with sensitivity extending into the near infrared, by L. R. Koller[30] of the
General Electric Laboratories in 1930; of the cesium-
antimonide photocathode, with maximum quantum efficiency
(at the blue end of the spectrum) of about 15 percent,

by P. Goerlich[31] in Germany in 1936; of the panchromatic
bismuth-oxy-cesium photocathode by A. H. Sommer[32] of
RCA in 1939; of the multialkali photocathodes with peak
quantum efficiencies up to 35 percent coupled with a
broad spectral response, also by A. H. Sommer,[33] in
1955; and, finally, of highly sensitive photocathodes
prepared by the deposition of cesium films on gallium
arsenide, gallium phosphide and related materials by a
number of research teams in the 'sixties.[34,35] These
developments were accompanied by a deepening of the
understanding of the factors determining the performance
of a photoemitter in terms of solid-state physics. In
fact, the last-mentioned materials were the first photocathodes which were _designed_ on the basis of known physical principles.

     B. _Secondary Emission_. As we shall see in the
following section, in most camera tubes employed in
electronic television an electron beam serves to complete
the circuit between one or a sequence of photosensitive
elements and the power source maintaining the difference
of potential between cathode and anode. It is often
convenient to think of the electron beam as simply replacing the electrons or negative charges which have
been lost by e.g. a photocathode as the result of
illumination. However, this picture is frequently inadequate. It was shown by Austin and Starke,[36] who
discovered the phenomenon of secondary electron emission
in 1902, that, under certain circumstances, electron
beams could eject from metallic surfaces electrons exceeding in number the incident beam electrons. Thus, in
the presence of an electrode positive with respect to
the target, which can collect the emitted electrons, the
electron beam may charge the target positive rather than
negative.

     Our present picture of secondary electron emission
is, briefly, the following.[37] As the beam electrons
enter the surface of the target, a few of them are
scattered elastically (i.e. without loss of energy) by
the positively charged atom cores of the target substance. These are commonly referred to as "reflected"
electrons. Some others penetrate a small distance into
the target, transferring a fraction of their energy to
the target electrons, and then are "back-scattered" out
of the surface. The remaining beam electrons penetrate
into the target, dissipating their energy until it has
been reduced to a value comparable to that of the target
electrons. In this process, some of the target electrons
are given sufficient energy to leave the target; not all

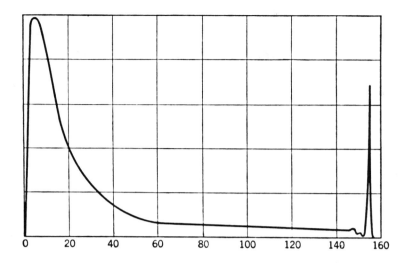

Fig. 8. Secondary Electron Velocity Distribution from
Gold (from McKay, reference 38, courtesy of Academic
Press, Inc., New York).

of them do so however -- in part because their direction
of motion does not favor this course, in part because
the electrons of higher energy tend to dissipate their
energy to unexcited target electrons on the way back to
the surface. However, a fraction of the excited target
electrons manage to escape. They are called "true
secondary electrons." Their mean energy of emission is
small, commonly of the order of 2 electron-volts. By
comparison, the reflected and backscattered electrons
have energies of the same order as the primary beam
electrons. (Fig. 8).[38]

     The preceding mechanism readily interprets the
characteristic yield curves for secondary emission shown
in Fig. 9. At very low energies of the bombarding
primary electrons, the latter can give the energy
necessary for escape to only very few target electrons,
so that the secondary-emission ratio (ratio of number
of emitted electrons to number of incident electrons)
is small. As the beam energy increases, so does the
secondary emission ratio. However, a maximum of the
secondary emission ratio is soon reached, since, for
higher beam energies, the rate (per unit distance) at
which the beam electrons transfer energy to the target
electrons is inversely proportional to their energy.
Beyond the maximum there is a gradual decline in the

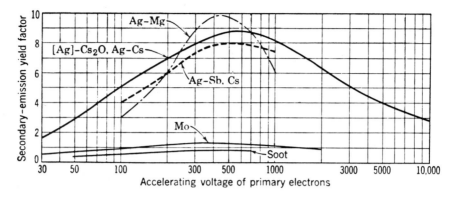

Fig. 9. Variation of Secondary Emission Ratio as
Function of Accelerating Voltage of Primary Electrons
(from V. K. Zworykin and G. A. Morton, "Television,"
Wiley, New York, 1954).

secondary emission ratio.

     Apart from the general shape of the yield curves
indicated in Fig. 9, there are wide variations depending
on the target material. For most of the materials encountered in vacuum tubes, the secondary emission ratio
reaches unity at a primary energy below 100 electron
volts and returns to unity at several thousand electron
volts. For pure light electropositive metals (which
cannot be maintained in this state in ordinary vacuum
tubes) the secondary emission ratio remains well below
unity throughout; for these materials the probability of
back scattering is small and the probability of energy
losses by excited target electrons on the way back to
the surface high, so that few manage to escape. On the
other hand very high secondary emission ratios are registered for certain semiconductors and insulators, in
which there are few electrons to which energy can be
transferred by excited target electrons with sufficient
energy to escape. The best photoemitters fit this
classification and possess a low work function facilitating the escape of the secondary electrons. Silver-
magnesium[39] and copper-beryllium targets, in which the
electropositive element is superficially oxidized and
the metal matrix facilitates resupply of the secondary
electrons, may also be regarded as insulator secondary
emitters. Very high gains are obtained from pure alkali
halides such as potassium chloride, although these are
usable only at low current levels.[40,41] The highest

gains -- 130 to 2500 electron volts primary energy --
have been realized with gallium phosphide targets with
an adsorbed cesium layer.[42]

Conversely, secondary emission can be minimized by
giving the target surface a porous structure (carbon,
nickel black etc.) which tends to trap escaping secondary electrons.

In most of the camera tubes now employed in electronic television, such as the image orthicon and
vidicon, a "low-velocity beam" is used to compensate
changes in target potential brought about by the light
distribution in the scene. This means that the bombarding energy is at all times below that for unity
secondary emission ratio, so that the beam tends to
drive the surface down to the potential of the beam
cathode. Secondary electrons are readily collected by
the anode which is generally several hundred volts
positive with respect to the beam cathode (and the target). (Fig. 10(a). In such tubes a low secondary-
emission ratio is obviously advantageous, since it
minimizes the beam current required to discharge a given
charge pattern. The image iconoscope constitutes an
exception, since here the secondary electrons generate
the picture signal.

A low-velocity beam could be employed only after the
problems of focusing and deflecting such a beam had been
solved. Focusing and deflecting a "high-velocity beam,"
bombarding the target with energies of several hundred
electron volts, was accomplished at a much earlier time
and utilized in earlier television camera tubes, such as
the iconoscope and image iconoscope. The beam now tends
to drive the target surface positive, to a potential
close to that of the collector potential; at equilibrium,
the number of electrons reaching the bombarded surface
element just equals the number leaving it, the excess
secondary electrons being returned to the scanned element by the retarding electric field in front of it.
The escaping electrons reach the collector only in part,
the remainder being distributed over the target, where
they tend to reduce the surface potential.[43] (Fig. 10(b).

Directly after scanning, the target element is
generally too positive to permit the escape of any photoelectrons (e.g. in the iconoscope) excited by illumination. This becomes possible only as the redistributed
electrons -- to which the photoelectrons eventually
contribute -- reduce the element potential. The effi-

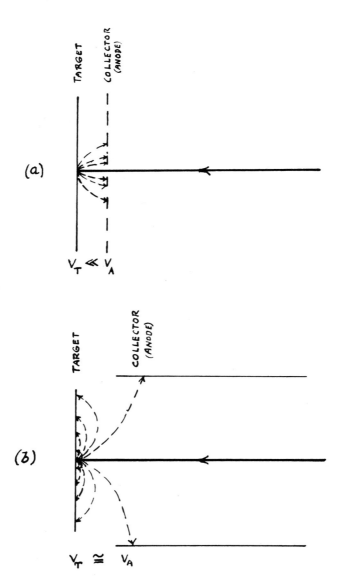

Fig. 10. Typical Behavior of Secondary Electrons Emitted
by Target in Response to (a) Low-Velocity Beam and (b)
High-Velocity Beam.

ciency of collection of photoelectrons as well as of secondary electrons is thus greatly reduced. Furthermore, the variation of the picture signal with illumination is far from linear -- a condition not necessarily unfavorable for picture reproduction. Finally the nonuniformity of redistribution over the target surface results in spurious shading patterns in the reproduced picture unless compensated by shading circuits.

Secondary emission plays one other very important role in television camera tubes -- that of signal amplification with a minimal addition of "amplifier noise."[44] Thus, in the image orthicon it is used twice for this purpose: First, in an "image stage" (Fig. 11 (a)), the image is converted into a charge storage pattern with an amplitude greater by a factor $(R-1)$ (R = secondary-emission ratio of the target for the accelerated photoelectrons) than that which could have been produced by photoemission directly; second, in a signal electron multiplier, the electrons leaving the target are multiplied, by successive secondary emission amplification, by a factor or the order of a thousand -- sufficient to override any noise sources in the succeeding thermionic amplifier.

The basic principle of secondary-emission multiplication is very simple (Fig. 11(b)). The primary electron beam is incident on a first secondary-emissive target or "dynode," the emission of the latter is guided to a second dynode, that from the second dynode to a third, etc., the electron current increasing at every step by the secondary emission ratio for electron energies corresponding to the potential difference between successive dynodes. The chief requirements are that the electrodes are so shaped and placed that the emission from the active area of one dynode is focused without loss onto the active area of the next dynode and that accelerating fields for the emitted electrons exist in front of these active areas. If the secondary emission ratio is much greater than unity the noise, or mean square fluctuation, in the output signal exceeds that obtained with a (hypothetical) noisefree amplifier by a factor $1 + 1/R_1$, R_1 being the secondary emission ratio of the first dynode.

The secondary emission multiplier, originally developed as a by-product of television research, has become a tool of major importance in all branches of natural science.

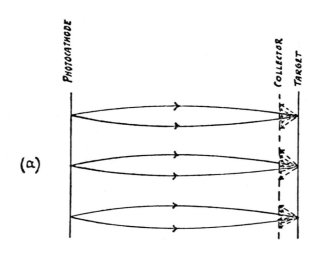

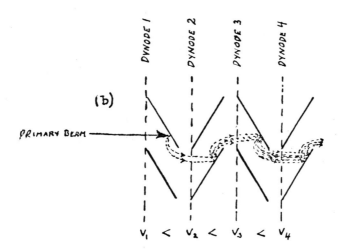

Fig. 11.   Principle of (a) Image Multiplication of
Charge Distribution and (b) Signal Multiplication by
Secondary Emission.

C. Electron Optics. The conversion of a visual
image -- a two-dimensional distribution of light and
shade -- into an electrical signal requires not merely
a photosensitive element which translates differences in
light intensity into differences in current or voltage,
but also a "commutator" which successively causes the
photoemission derived from different picture elements to
actuate a common signal generator, or, as an alternative,
successively derives the output signal from individual
photoelements associated with the picture elements.
Similarly, in picture reconstruction, a commutator is
needed to apply the received signal successively to
elements in the picture frame corresponding to the picture elements at the transmitter from which the signal
has originated.

In electronic television the commutators used for
both purposes are electron beams. In order that the
reproduced picture may be a faithful replica of the
original scene, these beams must be deflected in a
precisely controlled manner; to realize sharp, high-
quality pictures, they must be sharply converged.
Electric and magnetic field are the means used for
accomplishing both purposes.

The design of electric and magnetic fields to focus
and deflect electrons in a prescribed manner is commonly
called electron optics. The term follows from the recognition of William Rowan Hamilton, over one hundred
years ago, that the paths of material particles subject
to conservative force fields obey the same mathematical
laws as light rays in a medium of variable refractive
index. Much later, in 1927, Hans Busch[45] showed, both
theoretically and experimentally, that axially symmetric
electric and magnetic fields act indeed on electron
beams in the same manner as ordinary glass lenses act
on light beams. The "refractive index" n for electrons
in a field with electrostatic potential V and magnetic
vector potential A can be written simply

$$n = \sqrt{\frac{2eV}{mc^2}}\ \sqrt{1 + \frac{eV}{2mc^2}} - \frac{eA}{mc}\cos\theta \quad ,$$

where c is the velocity of light and θ the angle between
the path and the magnetic vector potential. The zero
level of the potential V is made such that eV represents
the kinetic energy of the electron. It is thus possible
to derive the path equations of the electrons from
Fermat's law of optics:

$$\int_A^B n\, ds = 0 \quad .$$

Fermat's law states that for the actual light ray (or electron path) from point A to point B the optical distance is a minimum or maximum as compared with any adjoining comparison path.

In any actual electron-optical system only the electrodes surrounding the region through which the electrons move, along with their potentials, as well as external current carrying coils and magnetic cores can be determined at will. The fields in the interior, which enter into the refractive-index expression and the path equations, must be derived from a solution of Laplace's equation for the boundary conditions established by the electrodes and magnetics. For electrostatic systems Laplace's equation is simply:

$$\nabla^2 V = 0 \quad .$$

The determination of electron paths within the system is thus normally carried out in two steps: the determination of the fields and the solution of the path equation in these fields. However, computer programs applicable for a great range of practical cases, have been written for carrying out both operations. With them, the computer supplies the electron paths if the point of origin and initial velocity of the electron as well the boundary potentials are specified.

The most common form of an electronic "commutator" used in television receiving and transmitting devices is shown schematically in Fig. 12. The source of the electrons is a small flat thermionic cathode, consisting of a nickel disc coated with a relatively stable low-work function material, such as a mixture of barium and strontium oxides. As the material is heated, a minute fraction of the electrons within it attains sufficient energy to overcome the potential barrier at the surface which prevents the bulk of the electrons from escaping. These electrons are accelerated toward a positively biased first anode and at the same time deflected toward the axis by the negatively biased control grid. They form a pencil of minimum cross section (the "crossover") at the point where the principal paths (i.e. the paths of electrons leaving the cathode with zero velocity) cross the axis and diverge from this point on toward a final electron lens, which serves to image the crossover

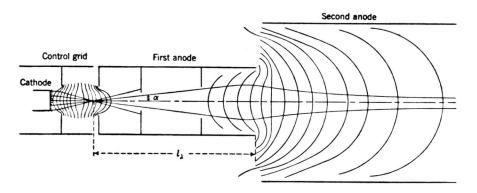

Fig. 12. Simple "Electron Gun" for Forming Electron
Beam Scanning Screen or Target (from V. K. Zworykin,
et al., "Electron Optics and the Electron Microscope,"
Wiley, New York, 1945).

on the image screen or target. The final lens illustrated is that formed between two cylinders at different
potentials; the sequence of curved lines represent the
equipotential surfaces which can be thought of as re-
fracting the electron paths in the same manner as a
boundary surface between two media of slightly different
refractive index. The magnetic field of a short solenoid
can be similarly employed for imaging the crossover on
the screen although the detailed interaction between
field and electron is quite different. The intensity
of the electron beam is varied by changing the potential
of the control grid; as the potential of the latter is
reduced a smaller fraction of the electrons emitted by
the cathode pass through the grid aperture, the remainder
being turned back toward the cathode. At the same time,
the crossover moves toward the cathode; this has,
however, only a minor effect on the sharpness of focus
on the screen since the crossover displacement is small
compared to the distance between the crossover and final
lens.

     The system shown in Fig. 12, designated as an
electron gun, serves merely to form a sharply focused
electron spot of controllable intensity at one point on
the screen or target. To effect the "commutation" the
beam is subjected to a pair of transverse magnetic (or
electric) fields just beyond the final lens. The exciting currents (or voltages) for the horizontal and
vertical deflector exhibit a sawtooth-shaped variation

with periods corresponding to the time required for
describing a single scanning line (about 60 microseconds) and a complete picture field (1/60 second) respectively, the electron emission being suppressed
during the short return times of the sawtooth. As a
result, the electron spot covers the screen and target
area with a closely spaced raster of horizontal lines;
the deflections at the transmitter and receiver are
synchronized so that the beam scanning a certain point
of the image area in the receiver is modulated by the
signal derived from the corresponding point of the scene
at the transmitter. The design of deflection systems
becomes a complex electron-optical problem when practical
considerations (such as large viewing screens and small
receiver depth) demand large angular ranges of deflection, commonly of the order of 110°.

     In several different types of television trans-
mitting devices another type of electron-optical imaging
is required; not simply the imaging of a point-like
crossover, but that of an extended area. The object
commonly used is a photocathode, on which an image of
the scene is projected. If the desired electron image
is comparable with the object in size the imaging is
effected most simply by an axial focusing coil or
solenoid surrounding the tube, producing an approximately
uniform magnetic field within it. This magnetic field
is superposed on an electric field accelerating the
electrons toward the anode and image plane. The electrons from any one point of the cathode spiral about the
magnetic field lines, reuniting in a common node at the
image plane for a proper choice of the magnetic field
intensity (Fig. 13). This system lends itself to directing the photoemission from successive picture elements
to a common signal generator. For this purpose horizontal and vertical magnetic deflecting fields are
superposed on the focusing and accelerating fields by
coils outside of the vacuum envelope and a small
aperture in front of an electron collector is placed at
the center of the image plane. The deflecting fields
cause the electron image to sweep across the aperture
so that, at any one time, electrons from just one picture element contribute to the output signal.

     D. Phosphors and Other Electron-to-Light Converters
When, in the picture reproducer, the modulated electron
beam has been directed to the appropriate element of
the image area, the problem presents itself of efficiently converting variations in beam current into
corresponding variations in picture brightness. The

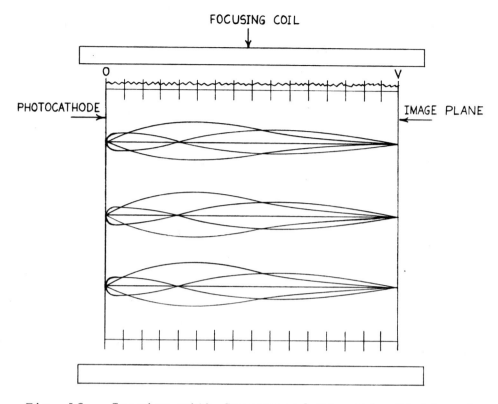

FOCUSING COIL

PHOTOCATHODE

IMAGE PLANE

Fig. 13.   Imaging with Superposed Magnetic Field and
Electric Accelerating Field.

target material either converts a portion of the beam
energy directly into light or is modified in its optical
properties by the beam so as to alter the amount of light
from an external source which is transmitted, reflected,
or scattered at the picture element.

        Direct conversion of electron energy into light
occurs at the screen of the ordinary television picture
tube or kinescope. The substance which accomplishes
the conversion is called a cathode-luminescent material
or, more simply, a phosphor.

        As H. W. Leverenz[46] points out, inorganic phosphors

which emit visible light after illumination with light
of shorter wavelength have been known since the beginning of the seventeenth century. Several naturally
occurring minerals have this property. These were
found to luminesce also under electron impact. Thus,
early cathode-ray tubes employed screens of powdered
willemite, a green-fluorescing zinc orthosilicate mineral
containing manganese as an impurity.

The demands of television for brighter displays
and black and white (rather than green) pictures were
an important impetus for the synthesis of phosphors
with properties superior to those of the naturally
occurring materials. In due time, these were supplemented by the requirements of the fluorescent lamp
industry. With the advent of color television, a need
arose for high-efficiency red and blue, as well as green,
phosphors. Finally, special applications, such as
flying-spot scanners employed as picture signal generators, required phosphors with decay times which were a
fraction of a microsecond.

Through painstaking research and exhaustive exploration of preparation techniques the chemist was able
to meet all these challenges. Even though the understanding of the cathodoluminescent process progressed as
more and more phosphor systems were investigated, the
approach remained to a large extent empirical and in-
tuitive. Practically all the important phosphors have
in common a stable, colorless, crystalline host lattice
containing impurities or activators varying in concentration from a few thousandths of a percent to a few
percent. For efficient energy conversion, the energy
of the electron beam absorbed by the host lattice must
be transferred in large part to the impurity centers;
furthermore, the excited impurity centers must return
to their normal state by the emission of visible radiation rather than by any one of a number of possible
competitive processes. The form and concentration of
the original ingredients; the fluxes employed in mixing
them; the atmosphere in which the mixture is heated;
the presence of traces of foreign substances; the heating and cooling cycle; and the mechanical method of
preparing the powder all effect the relative probabilities of these processes and hence the efficiency and
the spectral distribution of the light emission.

White phosphors in modern direct-viewing kinescopes
attain conversion efficiencies (i.e. ratios of the
radiant energy of the visual light output to the electron

beam energy incident on the phosphor) equal to about
0.1, making possible picture brightness of several
hundred footlamberts, comparable to that of an indoor
scene on a bright day.

Apart from the phosphors, there are substances
which darken or change color under electron beam impact.
These are the cathodochromics;[47] potassium chloride is
an early representative of this group.[48] In principle,
they permit utilizing diffuse ambient light for viewing
a picture traced by an electron beam. The picture is
normally erased by the application of heat. This process is too slow for practical use in television.

Another method of picture reproduction, which is
successfully employed in large-screen projection television, utilizes a thin low-vapor-pressure oil film on
a mirror surface. This is the Eidophor system, originally developed by H. Fischer in Switzerland.[49] A con-
stant-current scanning beam is employed, the picture
signal modulating a very high-frequency field component
superposed on the horizontal deflecting field. This
produces a ripple on the oil surface, with an amplitude
proportional to the picture signal. The oil film and
its mirror substrate form part of a "Schlieren" optical
system which, in the absense of oil-film distortions,
directs all the light from a concentrated arc source by
way of a slotted mirror back on to itself. As the oil
film is distorted by high-frequency ripple, light proportional to the ripple amplitude is diffracted or
scattered so as to pass through the mirror slots to a
projection lens forming an image of the oil film surface
on the viewing screen.

In essence, the Eidophor system provides a light
valve for the light flux from the arc lamp which is
locally controlled by deflection modulation of the
electron beam. There are a number of other physical
phenomena which have been utilized for the same purpose,
although they have not come into wide use up to the
present. One of these is the electro-optic light
valve.[50] An electrooptically active crystal plate,
placed between crossed polarizer and analyzer filters,
serves as image target. With one surface of the crystal
held at a fixed potential by a conducting film, the
other is scanned by the amplitude-modulated electron
beam; the beam charges the surface capacitatively,
establishing an electric field within the crystal which
effects a rotation of the plane of polarization of the
collimated polarized light beam passing through the

the crystal. The light transmitted by the analyzer,
which depends on the angle of rotation and hence on the
modulation amplitude of the beam, is utilized for
imaging the crystal target onto the viewing screen.

Suspensions of colloidal disk-shaped or rod-shaped
particles can also serve as light valves,[51] an electric
field changing the transmission or reflection of the
system by aligning the particles. The light scattering
induced in nematic liquid crystals by applied electric
fields is of special interest in view of the low power
requirements of this phenomenon.[52,53]

E. The Physical Basis of Color Vision. Up to this
point we have spoken of the scene to be transmitted as a
two-dimensional distribution of light and dark and thus
neglected the fact that natural scenes exhibit variations in color as well as in brightness. As a matter
of fact, television, in its early development, was
color-blind and this condition persisted, to all intents
and purposes, until the middle 'fifties. Long before
this time, black and white television had developed into
a major industry and a large fraction of the population
of the United States had a substantial investment in
black-and-white television receivers.

At the same time it was clear that television could
not fulfill its promise of vision at a distance without
including the aspect of color. The entrepreneur and
engineer as well as the public body exercising control
over frequency allocations and broadcast standards, the
Federal Communications Commission, faced a difficult
problem: It was not sufficient to devise a color television system which produced pictures of satisfactory
quality with equipment which could be provided for a
reasonable price; it was also necessary, that this new
color television system could be introduced in such a
manner that the investment of the public in black and
white sets and of the industry in studio equipment and
receiver coverage would be protected. The last criterion
could best be satisfied by meeting the criterion of
"compatibility." Compatibility meant that black and
white sets, without any modification, could receive
color broadcasts (in black and white, of course), and
that the new color sets conversely, could receive black-
and-white broadcasts as well as color broadcasts. With
compatibility, every set owner would have access to all
stations within broadcast range and every station access
to all receivers. There would be a gradual change-over
from black-and-white to color transmission and reception

as an increasing number of broadcasters recognized the
value of the added effectiveness given by color to their
message and an increasing number of television viewers
came to regard the enjoyment added by color well worth
the higher cost of the color receiver.

The extent of this treatise does not permit a detailed presentation of the manner in which compatibility
between color television and black-and-white television
was ultimately achieved. We shall merely indicate the
properties of human vision which were utilized to realize
this goal.

It has been known since Isaak Newton's time that
the color of an object was determined by the spectral
distribution of the light which it emitted. Thus if
information concerning the spectral distribution for
every picture element were transmitted along with its
brightness, this information could, in principle, be
used to reconstruct a faithful color image of the
original scene. However, color television based on this
principle would require an enormously expanded transmission channel and would be impractical for this reason
alone. Newton already recognized that many different
spectral distributions could produce the same color
impression; thus the white produced by sunlight could
equally well be produced by the mixture, in appropriate
proportions, of lights of any two complementary colors,
such as orange and cyan (greenish-blue) or purple and
green.

The requirements for reproducing a given color impression were made definite at the beginning of the
nineteenth century by Thomas Young,[54] who postulated that
all color sensations were composed of three "basic
sensations," red, green, and violet, mixed in different
proportions. Hermann von Helmholtz[55] interpreted Young's
theory half a century later by attributing color vision
to three independent processes in the nerve substance
of the retina. The possibility of matching any color
by mixtures of three appropriately chosen primaries was
clearly demonstrated by experiments of Helmholtz and
James Clerk Maxwell.[56] In more recent times, the Young-
Helmholtz theory of trichromatic vision has been given
direct support by the detection and the measurement of
the spectral response of three distinct photosensitive
materials in the cones of man[57] and other primates.[58]

The trichromatic theory suggested that it should
be possible to separate the color and brightness

distribution in the scene to be transmitted into its
three primary components (e.g., red, blue, and green),
to transmit these in succession over the same channel,
to reproduce them in the receiver, and to view them in
superposition. If the successive color fields followed
each other in a sufficiently rapid time sequence, per-
sistence of vision would cause them to merge into a
single image with the color and brightness distribution
of the original scene.

Standards corresponding to this simple scheme were,
in fact, temporarily adopted in the United States (in
1950-54);[59] but were soon abandoned; they failed to meet
the test of compatibility, apart from other shortcomings. In order to avoid intolerable flicker effects
in the image and to realize the desired merging of the
three color components, the field frequency had to be
increased from 60 per second, employed in monochrome
television, to 144 per second. At the same time, to
permit color broadcasting within the standard monochrome channel, the line number was reduced from 525
to 405. With these standards, color broadcasts produced a meaningless jumble on the screen of a standard
black-and-white receiver.

In order to achieve a compatible system, it was
necessary to utilize additional favorable properties of
human vision. One of these, explored by A. V. Bedford,[60]
is that the ability of the eye to resolve detail is much
less for detail which differs in color only, but not in
brightness, than for differences in brightness; furthermore, it is less in the color range between green and
purple than in the color range between orange and cyan.
A second property of the eye is that time variations of
intensity confined to very small areas are not perceived
as objectionable flicker even at frequencies low enough
that similar large-area variations would be intolerable.

A compatible color television system making optimal
use of these properties was developed at the RCA Laboratories.[61] Corresponding broadcasting standards were
worked out by the National Television System Committee
of the television industry and adopted by the Federal
Communications Commission in December 1953. These
standards govern color television broadcasting in the
United States today.[62]

In the present system two signals are generated;
one is the brightness of luminance signal, which is

essentially indistinguishable from the picture signal
transmitted by a black-and-white broadcast. The other
signal is the chrominance signal. Its amplitude indicates the saturation of color of the picture at any
point, ranging from zero for white or grey to a maximum for intense red, green, and blue, orange, or cyan,
and its phase the hue at the same point. The frequency
range of the chrominance signal is made much less than
that of the luminance signal, corresponding to the lower
acuity of the eye for color differences as compared with
brightness differences. After modulation of a carrier
close to the high-frequency end of the transmission
channel, the chrominance signal is added to the luminance
signal to form the complete transmitted video signal.
The exact carrier frequency is chosen so that the
amplitude of the modulated carrier reverses sign in
successive frames.

     A color television signal detected by a black-and-
white receiver reproduces the picture in normal fashion
by means of the luminance signal. The only effect of
the chrominance signal is the superposition on the
scanning lines of a fine dot pattern in regions of high
color saturation. This dot pattern becomes visible only
because of the non-linearity of the receiver and does
not detract from the quality of the picture.

     In a color receiver tuned to a color broadcast
the chrominance signal is automatically separated from
the luminance signal and processed so as to recover,
in combination with the luminance signal, the red, green
and blue component picture signals. These are applied
to the color viewing tube which reproduces, on its
screen, the three component pictures in superposition,
creating the visual impression of the original scene in
natural color. Actually, in correspondence with the
limited frequency range of the chrominance signal, the
finest detail is reproduced simply as a variation in
luminance; full color reproduction is reserved for the
range in which the eye can resolve color differences
fully.

     With a black-and-white signal applied to a color
receiver, a "color-killer" deactivates the chrominance
signal circuits. The same luminance signal determines
the intensity variation in the three component images
which, in superposition, reproduce the picture in black
and white.

     Thus the criterion of compatibility is fully

satisfied.

Unquestionably, numerous other aspects of television practice could be used to illustrate its indebt-
edness to scientific research. The preceding discussion
has limited itself to some of the key features of picture
transmission and reproduction.

CHAPTER III

Television Pickup Systems

A. <u>Storage and Non-Storage Systems: Iconoscope
and Image Dissector</u>. We have already noted, in Chapter
I, two basic types of television pickup systems --
storage systems and non-storage systems. In storage
systems the light emitted from any one picture element
is utilized continuously for the generation of the picture signal. Charge transferred by photoemission or
photoconduction is stored in a condenser and released
at the instant of scan to generate the signal. In non-
storage systems, on the other hand, only the light
emitted by the picture element at the moment of scanning contributes to the picture signal for the element
in question.

The application of the storage principle in the
television pickup has several advantages. The most important of these is greatly increased sensitivity. An
ideal storage system and an ideal non-storage system
differ in sensitivity by the number of picture elements
in the transmitted scene, or several hundred thousand
for current television standards. A second advantage of
the storage principle is that it permits the use of
photosensitive materials with relatively slow response.
Whereas with non-storage systems the reproduced picture
is smeared in the scanning direction if the photocurrent
produced by a light pulse persists longer than the scanning time for a picture element (a fraction of a microsecond), picture smear is observed with storage systems
only if the photocurrent persists longer than a frame
time, or 1/30 second; at that, unless the television
camera is panned, this smear will be observed only on
moving objects. Thus, application of the storage
principle permits the use of relatively slow, but
highly sensitive, photoconductive materials in television camera tubes. Finally, the storage principle
facilitates scan conversion: television images reproduced on the target of a storage camera tube can be
read off at a different scan rate, permitting the re-
broadcasting of television programs originating in an
area with different television standards.

The first practical storage camera tube, the
iconoscope perfected by the senior author in 1931, has
already been described. It utilized a "mosaic" of
minute oxidized and cesium-activated silver globules

deposited on a mica plate as a photosensitive target.
Every globule formed a minute storage condenser with the
continuous signal electrode deposited on the other side
of the mica plate. The photosensitive target was
simultaneously illuminated by the image of the scene to
be transmitted and scanned by a sharply-focused high-
velocity electron beam. In principle, the minute storage
condensers were charged by photoemission throughout the
period between successive scans and then discharged by
the high-velocity beam, returning the potential independent of illumination. We have seen, however, that,
with high-velocity scanning, the process is, in fact,
much more complex than this.[63] The equilibrium potential
under the beam is actually slightly positive with respect
to the collector electrodes for secondary and photoemission from the target, and the mosaic potential at
no time departs by more than a few volts from the
collector potential. Under these circumstances a large
fraction of the photoemission and secondary emission
from any one mosaic element is returned to the target
and, in part, redistributed to other elements (Fig. 14).
While this redistribution is essential to the operation
of the tube, it also results in a reduction in the picture signal amplitude to approximately 5 percent of that
achievable with an ideal storage system and in the superposition of spurious shading on the image. Finally, the
noise, or statistical signal fluctuations, contributed
by the thermionic camera signal amplifier exceeds that
of the iconoscope output signal, resulting in a further
reduction in the effective sensitivity of the system.

The non-storage counterpart to the iconoscope was
the image dissector, introduced by Farnsworth a few
years after the perfection of the iconoscope. Here the
image of the scene to be transmitted was projected on a
continuous photocathode and the photoemission from the
latter accelerated and reimaged by a uniform magnetic
field in the plane of a collector aperture. Transverse
magnetic fields superimposed on the constant focusing
field caused the electron image to sweep across the
collector aperture in a standard scanning pattern. In
later models of the dissector, the electrons passing
through the aperture were amplified by a secondary-
emission multiplier, overriding noise sources in the
camera amplifier. The noise level in the reproduced
picture was then set primarily by the irreducible shot
noise of the photoelectrons selected by the limiting
aperture.

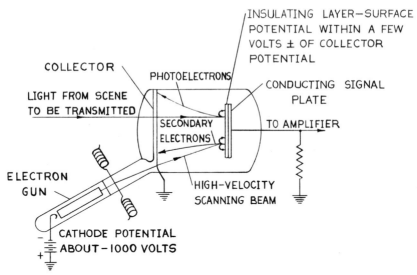

Fig. 14.   Iconoscope.

In early tests the image dissector, while inferior
to the iconoscope in sensitivity, produced excellent
images, substantially free from spurious shading, at
sufficiently high object illuminations. However, the
disparity between the non-storage image dissector and
the storage camera tubes increased rapidly with time:
On the one hand, the need for storage increased with the
increase in required picture resolution or line number.
On the other hand, whereas the image dissector with
electron multiplier represented an end-point in the
performance of non-storage systems, the iconoscope was
only a first step in the development of storage camera
tubes.   Even so, it relegated the image dissector and
mechanical non-storage pickup systems to a limited role
in film and slide transmissions and special-purpose non-
broadcast applications, where illumination intensity was
no problem.

     B.   Perfection of Storage Systems: Image Icono-
scope, Orthicon, Image Orthicon, and Image Isocon.
After the perfection of the iconoscope, an intensive
research program was undertaken to overcome its shortcomings and to design a camera tube approaching an ideal
storage pickup system in performance.   The first important step in this sequence was the image iconoscope
(Fig. 15).[64]   Here the optical image was projected not
on the mosaic, but on a continuous semitransparent

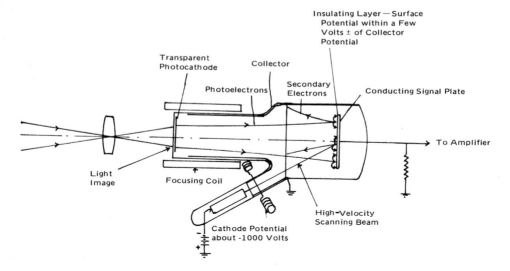

Fig. 15.   Image Iconoscope.

photocathode; the photoemission was then accelerated and imaged by a magnetic or electrostatic field onto the mosaic, which was scanned by a high-velocity beam as in the iconoscope. The greater photosensitivity of the photocathode and the secondary emission gain at the mosaic resulted in an overall gain in sensitivity of the order of 10. Various independently developed modifications of the image iconoscope were employed for many years as broadcast camera tubes in England and on the European continent.

A more radical innovation in design was realized by Rose and Iams[65] in the orthicon, which replaced the high-velocity scanning beam in the iconoscope with a low-velocity scanning beam (Fig. 16). The equilibrium potential of the target under the beam is here the cathode potential of the scanning-beam gun, so that both photoemission and secondary emission from the mosaic are collected by the anode structure without any distribution. This results in a gain in collection efficiency over the iconoscope by a factor of the order of 20, and the elimination of spurious redistribution shading. At the same time it required an axially symmetric structure, with projection of the optical image onto the mosaic through a transparent signal

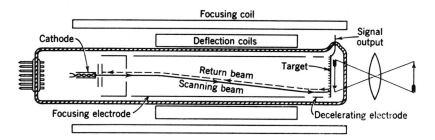

Fig. 16.   Orthicon-Type Pickup Tube.

plate, and novel methods of beam focusing and scanning
which assured orthogonal landing of the beam on the
target.   Focusing was accomplished by a long magnetic
field formed by a solenoid enveloping the tube, deflection, by superposed electrostatic and magnetic fields.
In more recent orthicons all-magnetic deflection has
been used; the electrons spiral about the resultant
magnetic field lines with an amplitude which is minimized by careful adjustment of the length and position
of the deflecting fields with respect to the beam nodes
formed by the focusing field.

        While the orthicon itself has not been widely em-
ployed as a camera tube, its successor, the image
orthicon developed by Rose, Weimer, and Law[66] (Fig. 17),
dominated television broadcasting in the United States
during the twenty-year period following World War II.
In the image orthicon an image section similar to that
of the image iconoscope is placed ahead of the target
so as to increase the amplitude of the stored charge
image and a secondary-emission electron multiplier
surrounds the gun structure to permit reduction of the
camera amplifier gain to the point that the amplifier
noise contribution becomes negligible.   The last arrangement is made possible by the fact that, with magnetic
orthicon scan and focus, the electrons which are reflected by the target as well as the ejected secondary
electrons, spiral about the field lines back toward the
gun aperture; thus a diaphragm surrounding the incident
beam can serve as first dynode for an axially symmetric
pinwheel secondary-emission multiplier structure.

        The image-orthicon target is a film of slightly
conducting glass stretched on a frame; a closely-spaced

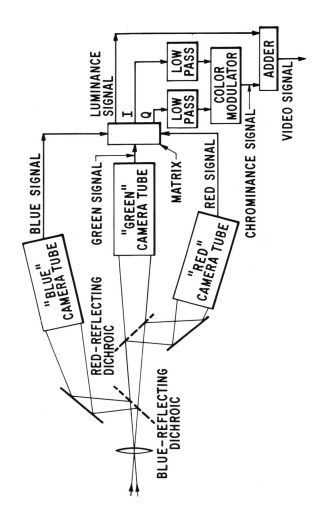

Fig. 17. The Image Orthicon.

fine-mesh electrode held at a potential slightly positive with respect to the gun cathode serves as secondary electron collector and prevents the target from rising to a potential for which the secondary emission for the scanning beam exceeds unity; if the latter happens, as it may when a light source is imaged on the target in the simple orthicon, the charge image on the target becomes unstable. The image orthicon, on the other hand, remains stable under all circumstances and rivals the human eye in sensitivity down to light levels of the order of full moonlight. For special purposes the sensitivity of the image orthicon can be enhanced by the addition of further image stages and/or special targets with high secondary emission to the point that the contribution of individual photoelectrons can be recognized in the image.

A final step in the perfection of a storage camera tube was taken by Weimer[67] with the image isocon. Here, by means of a carefully adjusted system of deflection fields and apertures, the secondary electrons from the target are separated from the reflected return electrons and only the secondary electrons are multiplied to form the output signal. In this manner the noise in the low lights of the image can be made to approach zero, whereas in the image orthicon it approaches a limiting value determined by the shot noise of the scanning beam. The relative complexity of construction and adjustment has prevented the use of the image isocon in television broadcasting.

C. Photoconductive Camera Tubes: The Vidicon. The vidicon was developed by Weimer, Forgue, and Goodrich in 1949 (Fig. 18).[68] Like the orthicon, the standard vidicon employs low-velocity scanning and superposed magnetic focusing and deflecting fields. However, the target is, instead of a photoemissive mosaic, a thin layer of photoconductive material deposited on a transparent signal plate. The signal plate is given a positive bias with respect to the beam cathode, whereas the layer surface scanned by the beam is driven by the latter to cathode potential. In darkness the layer is essentially an insulator, so that the bias voltage is maintained across it. Under illumination, the layer becomes conducting, causing the scanned surface to assume an increasingly positive potential between scans. The scanning beam deposits sufficient negative charge on the surface to return it to cathode potential and induces a corresponding signal current in the signal plate lead.

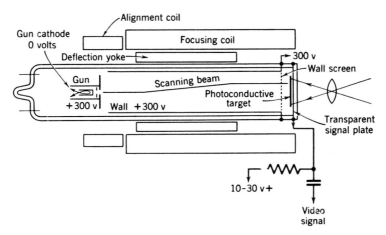

Fig. 18. The "Vidicon" Photoconductive Pickup Tube.

The vidicon, thus, fully utilizes the storage
effect. Furthermore, considerably higher sensitivities
(charge transferred per unit of incident light) may be
realized with photoconductive materials than with photoemitters, so that adequate sensitivity for most purposes
can be realized without the addition of an image section
and signal multiplication. Normally about 1 inch in
diameter and 6 inches in length, the vidicon is compact
as well as simple. It quickly became the camera tube
for the simpler camera tubes employed in non-broadcast
applications of television. The standard vidicon, with
an antimony-trisulfide photoconductive layer, remained
inferior to the image orthicon not only in ultimate
sensitivity, but also with respect to "lag," resulting
in a smear of moving objects transmitted at low light
levels, and in uniformity of response over the picture
area. The latter defects were overcome in vidicon-type
tubes with lead-oxide targets, designated as "Plumbi-
cons," developed at the Philips Research Laboratories
in Holland.[69] Lead-oxide vidicons have found application, particularly in studio cameras for color broadcasting, where their linear response and response uniformity over the picture area are particularly desirable.

Another important modification of the vidicon is the
silicon vidicon, using a target with some 600,000 individual silicon diodes per square centimeter formed on
a thin silicon wafer by integrated-circuit techniques.[70]
They are marked by a broad spectral response and immunity
to damage by overexposure and hence well adapted to the

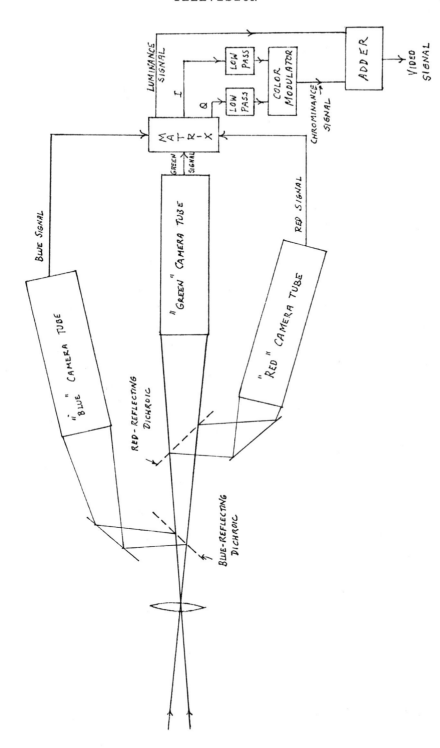

Fig. 19. Principle of Color Television Camera.

"Picturephone" service of the Bell System. The addition
of an image section to such a tube leads to the silicon
intensifier target camera with target gains up to 3000.[71]
This is a camera tube appropriate for use at extremely low
light levels.

Vidicons have also been constructed which combine
high sensitivity, extremely high resolution, and long
image retention, which make them particularly suitable
for e.g., the cameras of the Earth Resources Technology
Satellite. An example is the 4.5 inch return beam vidicon
with antimony sulphide oxy-sulphide (ASOS) photoconductor,
with a measured resolution of 6000 television lines.[72]

D. <u>Color Camera Systems</u>. The basic system of a
color television camera is illustrated in Fig. 19.[73] The
image of the scene to be transmitted is focused on the
targets on three camera tubes through a beam-splitter
system which separates the incident light into its red,
green, and blue components. The means for accomplishing
this are spectrally selective interference filters,
supplemented by trimming filters to realize a prescribed
spectral response for each of the three channels. The
spectral responses are selected so that, within the limits
of physical realizability, the three output signals,
applied in the picture reproducer to trace out red, green,
and blue component pictures in superposition, regenerate
the color distribution in the original scene.[74,75]

The camera objective commonly is a zoom lens with a
large range of focal lengths and corresponding field
angles. The back focus is made considerably larger than
the shortest objective focal length so as to accomodate
the beam-splitting system. The latter itself may be in
the form of a compact filter block, the lens correction
taking account of the fact that the ray paths beyond the
objective are largely in glass.[76]

In the earlier color television cameras, three image
orthicon tubes were employed for the three color channels.
In these cameras the optical systems as well as the scanning rasters in the individual tubes had to be aligned
very accurately to prevent misregistration of the red,
green, and blue component images. An improvement in the
alignment tolerances was realized in the four-tube camera,
in which the luminance signal (describing the brightness
variation in the picutre) was derived from a single high-
resolution image orthicon and the chrominance signal
(describing the hue and saturation of the color in the
picture) from a set of three lower-resolution vidicons

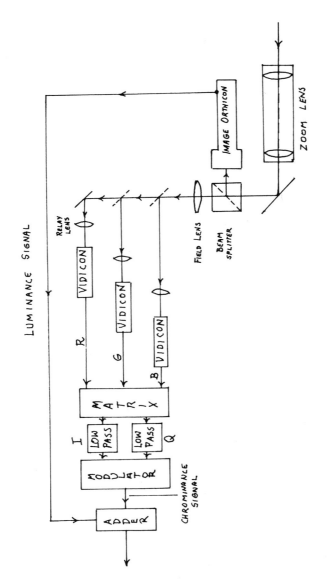

Fig. 20.   Four-Tube Color Television Camera.

(Fig. 20).[77] This arrangement utilizes the fact that the resolution of the human eye for color differences is much less than that for brightness differences.

The linear response and uniformity of sensitivity over the target area of the lead-oxide vidicon or "Plumbicon" made it particularly suitable for color cameras, facilitating faithful color reproduction. The alignment tolerances required in three-tube "Plumbicon" cameras have been reduced by utilizing the high-frequency components in the intensity contribution of the green component picture alone and adding the corresponding signal components to the red and blue signal outputs ("contours from green").[78]

In the four-tube camera the objective forms an image of the scene in the plane of a field lens, and this image, in turn, is reimaged by relay lenses onto the tube targets. The use of field and relay lenses makes it possible to accommodate a relatively bulky beam-splitting system without placing restrictions on the objective focal length. If suitable composite color gratings, with periods smaller than the smallest separation to be resolved in the image, are placed in the image plane near the field lens, a color camera with a single high-resolution vidicon may be realized (Fig. 21)[79] if the absorbing stripes of one of the grating components absorb only the red component of the image, the absorbing stripes of the other grating component (with the longer horizontal period) only the blue component, the output of the camera contains, apart from a low-frequency luminance signal, the red and blue picture signals modulating two high-frequency carriers. The red and blue signals may be recovered by envelope detection and the green signal obtained by subtraction of the red and blue signals from the luminance signal in suitable proportion.

There are many possible variations in the form of the gratings and the manner of color signal recovery. Furthermore, if the gratings are deposited in contact with the light-sensitive target, the relay lens system becomes unnecessary. While a number of practical difficulties have so far prevented such single-tube color cameras from being competitive in high-quality broadcasting, they have the obvious advantages of great compactness and complete absence of registration difficulties.

E. <u>Solid-State Cameras</u>. All television camera systems in current use (1970) employ electron beam scanning for converting the two-dimensional brightness distribution in the picture and its gradual change with

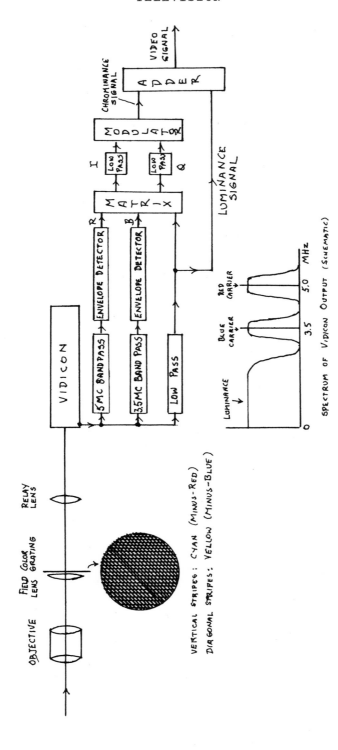

Fig. 21. Principle of a Single-Tube Color Camera.

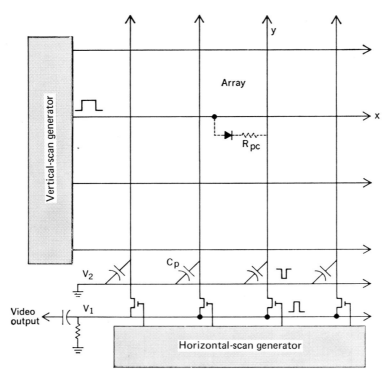

Fig. 22. Solid-State Image Sensor Using an Array of
Thin-Film Photoconductor and Diodes.

time into a one-dimensional time-varying electrical
signal. However, considerable progress has been made
in the construction, by integrated-circuit fabrication
techniques, of image sensors with solid-state digital
scanning circuits. Such sensors have the advantage of
extreme compactness, minimal power consumption, and,
in view of the absence of a vacuum chamber and thermionic
cathode, potentially indefinite life.

     The sensor consists of an array of elementary units
at least equal in number of the number of required picture elements. Every such unit incorporates a photosensitive element in the form of a photodiode, photo-
transistor, or photoconductor, a switching element
generally in the form of a diode, and provision for
element storage, commonly in the form of a storage
capacitor. Vertical and horizontal metallic leads
connect the individual elements to the horizontal and
vertical scan generators. The entire array, including

the circuit elements of the scan generators, is prepared
by a sequence of evaporations through appropriate masks.

Image sensors with a variety of circuit configurations have been prepared both on silicon chips and by
thin-film techniques, using CdS-CdSe, CdSe, and Te as
active materials.[80] With the latter, Weimer and his
associates have prepared sensor arrays of 256 x 256
elements, approximately half an inch on the side, capable
of being scanned in 1/60 second at an elemental scan
rate of 4.8 MHz. The circuit employed is shown schematically in Fig. 22. When a vertical pulse renders the
diodes of a particular row conducting, the column
capacitors Cp are charged through the photoconductor
R_{pc}, thus providing line storage in addition to the
intrinsic "excitation storage" (i.e. the build-up of
conductivity over a period of the order of the frame
time) of the high-gain photoconductor. The stored
charge is delivered to the video output as the horizontal
scanning pulse opens the corresponding transistor gate.
Both horizontal and vertical scanning is effected by
pairs of 16-stage shift registers operating in series,
so that only 64 external leads are needed to scan the
65,536 elements of the array. Apart from residual
vertical streaking owing to array nonuniformities, the
solid-state camera transmits satisfactory 200-line pictures at ordinary laboratory illumination.

CHAPTER IV

Reproduction of Television Pictures

A. The Kinescope. The essential elements of
today's black and white picture tube were incorporated
in the "kinescope" presented to the public in 1929.
Subsequent changes were mainly in the nature of refine-
ments.[81] The electron gun forming the scanning beam
underwent various modifications designed to permit
operation at higher voltages and to deliver smaller spot
sizes less subject to "blooming" with current increase
and deflection over large screens; the electromagnetic
deflection yokes extended the deflection angles of the
beam over a range of 110° or more, permitting large
ratios of the screen diameter to the tube length, and
hence "thin" television sets, easily fitted into modern
furnishings. The screen itself came to be prepared with
highly efficient composite zinc-sulfide phosphors, delivering a black-and-white image, in place of the green
image reproduced on the natural willemite screen of the
original kinescope. The deposition of a thin-aluminum
film on the back of the screen increased picture brightness, and eliminated charging effects and screen damage
from ion bombardment. Finally, filter-glass face plates
improved contrast in the presence of ambient illumination
so as to permit satisfactory viewing in a moderately
well-lighted room.

B. Color Viewing Tubes. A new challenge was presented, however, by the problem of reproducing pictures
in color on the screen of a picture tube. An intensive
exploration of different approaches to this problem at
the RCA Laboratories[82] in Princeton, N.J. in the late
1940's and early 1950's resulted in the selection of the
three-gun shadow mask tube, based on early proposals of
Arthur N. Goldsmith and Alfred C. Schroeder and perfected
by Harold Law,[83] as the most promising design. Three
electron beams, modulated by the red, green, and blue
picture signals, respectively, are here aimed at the
center of the phosphor screen, forming angles of the
order of 1° with respect to the tube axis, and deflected
jointly over the screen area (Fig. 23). The shadow
mask, a metal screen perforated with a hexagonal array
of small round apertures with a center-to-center spacing
of e.g. 0.028 inch, is placed approximately 0.5 inch in
front of the screen, which is deposited on the inside
wall of the faceplate. The screen itself is an array of
red, green, and blue phosphor dots, one set for every

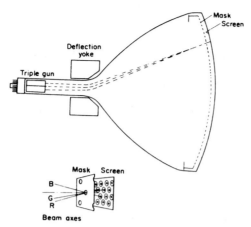

Fig. 23. Schematic diagram of RCA shadow-mask color
viewing tube.

shadow-mask aperture, so placed that the portion of the
red beam transmitted by the shadow-mask apertures strikes
only the red phosphor dots, that of the green beam the
green phosphor dots, and that of the blue beam the blue
phosphor dots. The dots are formed on the face plate
with a precision of approximately 0.001 inch by a photographic technique, with light sources at the beam deflection centers and the shadow-mask in place. A
complex asymmetric lens is employed to simulate the shift
of the deflection centers with deflection angle.

     After its commercial introduction in 1953, the
shadow-mask tube was continuously improved by advances
in manufacturing technique and design. New phosphor
combinations were introduced, to realize higher light
output and better color balance. Thus RCA presently
replaced the original combination of zinc phosphate red,
zinc silicate green, and mixed calcium-magnesium silicate
and zinc sulfide blue by a more efficient combination of
sulfide phosphors. A further significant step was the
wide-spread adoption of highly efficient red rare-earth
phosphors, initiated by the announcement by Sylvania,
in 1964, of an yttrium-vanadate: europium screen
material. A further improvement of screen brightness
was made possible by surrounding the phosphor dots by a
black matrix, absorbing incident ambient light and
permitting the use of a lower-density filter face plate.

     It is natural that, with the expanding demand for
color receivers, the search for new ways of reconstruct-

ing the color picture has continued. Primarily two
aims have been pursued in devising alternatives to the
conventional shadow-mask tube: to realize greater
screen brightness by minimizing or avoiding altogether
the beam current interception by the shadow-mask and to
simplify the tube, e.g. by one-gun instead of three-
gun operation.

    The conventional shadow-mask intercepts at least
70 percent and, in early tubes, intercepted as much as
88 percent of the incident beam current. This loss is
avoided altogether in various proposed tube designs in
which a three-color phosphor line pattern on the screen
is supplemented by stripes of secondary-emitting
material.[84],[85] The secondary emission of these stripes
in response to bombardment by the scanning beam (or a
"pilot beam" accompanying the scanning beam) excites
correcting signals applied to the beam deflection which
assure that the scanning beam (or beam trio) is at all
times properly aligned with the phosphor lines on the
screen. Unfortunately, the necessity of confining the
scanning beam to a width less than the width of the
phosphor lines on the screen can readily result in a
reduction in beam current exceeding that effected by
shadow-mask interception.

    The beam-current interception by the shadow-mask
is greatly reduced without requiring a reduction in the
size of the scanning spot if the shadow-mask function
is exercised by a grill of thin parallel wires, geo-
metrically registered with a phosphor line pattern on
the screen; if a potential difference two or three
times the accelerating potential of the incident beam
is applied between the wire grill and the screen, the
beam electrons are focused into narrow line segments
centered on the phosphor lines (Fig. 24).[86] Color
viewing tubes based on this principle are manufactured
by the Sony Corporation and represent the only significant departure from the conventional shadow-mask design
in tubes available commercially in 1970. If an
alternating voltage is applied between interleaved sets
of adjoining grill wires, the line segments formed at
the center of the green phosphor lines by a single
scanning beam can be shifted alternately to the adjoining red and blue phosphor lines (Fig. 25). The resulting
"single-gun Chromatron" is commonly associated with name
of Nobel Laureate Ernest O. Lawrence, who devoted much
effort to its perfection in the early 1950's.[87] The
deflection principle utilized is due to A. C.
Schroeder.[88]

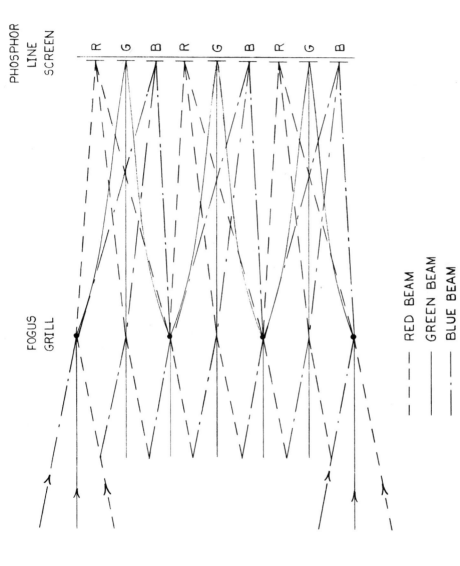

PHOSPHOR
LINE
SCREEN

R G B R G B R G B

FOCUS
GRILL

---- RED BEAM
—— GREEN BEAM
—·— BLUE BEAM

Fig. 24. Focusing and Color Selection in the Three-Beam Focus Grill Tube.

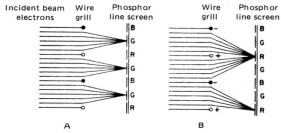

Fig. 25. Focusing and Color Selection in the Single-
Beam Focus Grill Tube. A. Grill wires at uniform
potential (green display). B. Alternate grill wires
at different potential (red display).

Significant improvements have been devised in the
design of focusing-grill tubes.[89] However, at the time
of writing (1970), these and the many other proposals
for color display have not detracted from the dominance
of the conventional shadow-mask tube. One-gun color
tubes suffer from the inherent drawback that the beam
can dwell on any one phosphor at best one third of the
time, leading to a loss in screen brightness which
generally exceeds that resulting from beam interception
by the shadow mask.

C. _Projection Systems_. Theatre television projection systems capable of providing pictures of acceptable brightness and definition on a 15 x 20-foot screen
were demonstrated as early as 1941.[90] They employed
relatively small kinescopes (with 7-inch screens)
operating at up to 80 kilovolts with an average beam
current of 2 milliamperes. The highly efficient projection system consisted of a spherical mirror with an
aspheric correction plate in the plane of its center of
curvature, resembling a Schmidt astronomical telescope
(Fig. 26). More recently (since 1949) the combination
of three similar units of smaller size, with red, green,
and blue phosphor-screen kinescopes, has been widely
employed for the projection of color television pictures
on medium-size screens.[91] A quite different and much
more complex, but highly successful, theater television
system is the Eidophor system, originally developed by
F. Fischer in Switzerland (Fig. 27).[92] In place of the
usual phosphor screen, this uses a thin-oil film,
illuminated by collimated light from an intense arc
source. The scanning beam modulates the surface of the
oil film, so that it scatters a portion of the incident
light proportional to the picture signal. Only the

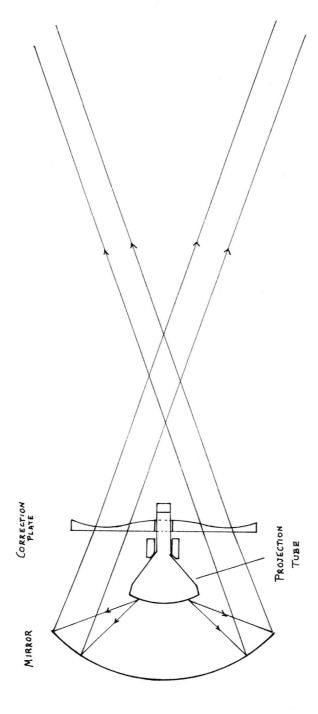

Fig. 26.   Television Projector.

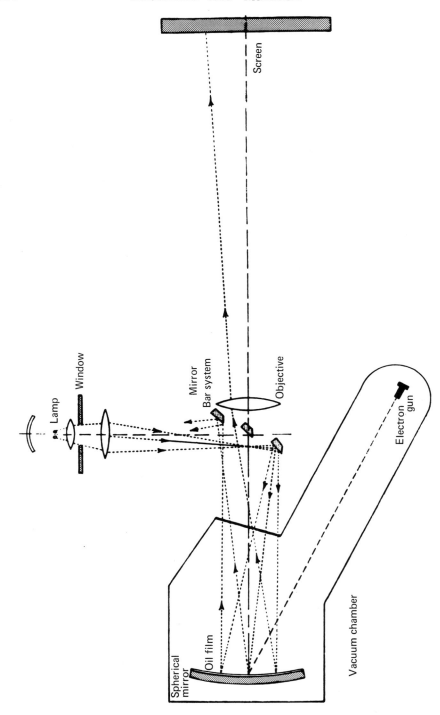

Fig. 27. Diagram showing the principle of operation of the Eidophor.

scattered light is utilized for forming the projected
image, the remainder being intercepted by appropriately
placed stops in the optical system. An ingenious extension of the Eidophor principle is realized in a
light-valve color television projector recently described
by W. E. Good of the General Electric Company.[93] Here
the electron beam generates complex grating patterns on
the oil surface which permit the reproduction of color
as well as brightness variations in the reproduced
picture.

The most recent approach to projection television
is the mechanical scanning of the picture screen by
three optically superposed individually modulated red,
green, and blue laser beams. Such a system has been
developed by Y. Yamada of Hitachi for the 1970 world
exposition in Japan.[94] The results are said to be
excellent. In view of the high laser powers required,
the system can scarcely be expected to be economically
competitive at the present time.

D. Panel Television Displays. The construction of
flat television displays dates from early days of television development. Thus Karolus, in 1934-35, built a
panel of 10,000 individual tungsten filament lamps with
the picture signals from 100 picture elements in a
horizontal line. The picture signal was applied simultaneously over as many channels to the lamps in one row;
mechanical commutators effected the vertical deflection,
or the shift of the signal from the lamps in one row to
those in the row below it.[95] The primary aim here was
to achieve much higher brightnesses in a large display
than could be realized at that time by projection
techniques. The advent of solid-state circuit elements
and integrated-circuit technology stimulated the hope
that flat displays of the "picture on the wall" type
might presently replace the conventional television
receiver console in the home as well. A way of
achieving this, which is simple in concept, would provide for the application of horizontal and vertical
gating pulses to a set of gates connected to vertical
and horizontal crossbars. These gates would control the
application of the video signal to light transducers
located at the crossbar intersections. (Fig. 28).

Electroluminescent phosphors,[96] which emit light
when an alternating voltage is applied to a thin phosphor layer sandwiched between e.g. an opaque and a
transparent electrode, appeared particularly suitable
as light transducers. Light emitting diodes, which had

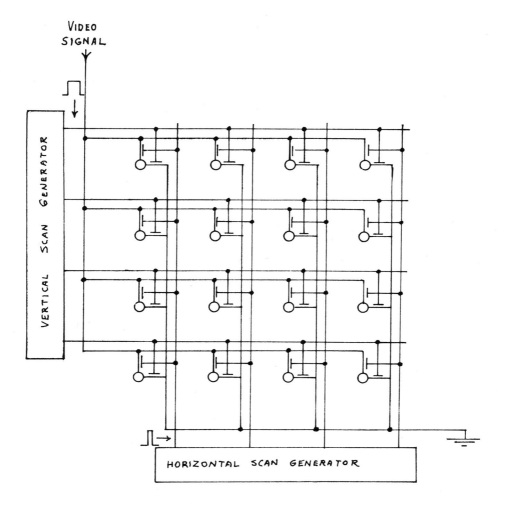

Fig. 28.   Principle of Television Panel Display.

If the circles represent simple light transducers, this
is a display without storage. If they represent elements
(such as transfluxors or ferroelectric capacitances)
which control continuously the power supplied to the
light transducer and are set by the gate video signal,
the display has frame storage.

the advantage of requiring lower operating voltages,[97] and a.c. gas discharges[98] in gases at low pressure between insulator-coated electrodes also became contenders. All of these, along with circuit elements such as transistors, diodes, resistances, and capacitances lend themselves to panel construction and the eventual manufacture of the panel by automatic machinery.

However, the simple system described, even if perfectly executed, would not provide an acceptable panel display. Since the light transducer is turned on only for element time, or about 0.1 microsecond out of a frame time of 30 milliseconds, the panel brightness is entirely inadequate. It is essential that the video control signal be stored for a time much larger than element time -- for frame time, if possible. This can be accomplished, e.g., by coupling the individual electroluminescent cells to the common a.c. power supply through transfluxors, minute transformers whose coupling coefficient is set by the video signal at the instant at which the vertical and the horizontal scanning pulses coincide at the element in question.[99] As an alternative, line storage can be achieved by using a horizontal scanning pulse traveling along a magnetostrictive delay line to store the video signal for a line period in the successive elements of a video register feeding the matrix columns.[100]

One of the simplest panel display systems suggested and demonstrated uses simply an electroluminescent layer in series with a nonlinear resistance layer (such as CdS powder in a plastic binder) and a piezoelectric layer sandwiched between two plane electrodes, to which the alternating voltage modulated by the video signal is applied.[101] Acoustic transducers on two sides of the display launch pulses in the piezoelectric layer at line frequency; the traveling intersection of the two pulse waves provides the gating pulse for the excitation of the electroluminescent layer and thus forms the scanning lines of the display, which are displaced relative to each other by appropriate phasing of the pulses. Another interesting approach, the "Elf" screen,[102] utilizes ferroelectric (barium-strontium titanate) condensers in series with the individual electroluminescent cell of an array. The reactance of the individual ferroelectric cell is "set" by the video signal in conjunction with the coinciding scanning impulses as described for the system with the transfluxors.

Even with all the suggested solutions, the technical and economic problems of making panel displays for television standards remain formidable. It is perhaps significant that perhaps the most ambitious effort in this direction, reported by Shimada of the Sony Corporation in 1968,[103] returns to the miniature tungsten lamp as light source. 78,000 such lamps, with red, green, and blue filter caps, are arranged in a panel with an 8-foot diagonal, to reproduce television pictures in color. While the total number of color trios is by an order of magnitude less than in a normal, color television display, the total absence of blooming and of registration errors leads to a relatively good picture. The scanning system uses line storage; horizontal pulses traveling along a delay line sample the video signal and the samples, applied to each column of the matrix, are converted into pulses of uniform amplitude and a width proportional to the video signal amplitude, ranging from zero to a line period. These pulses light the lamps in the row to which silicon-controlled rectifiers, actuated by the vertical scanning pulse, supply power. The requirements in electronic components of such a system -- not to mention the power requirements (10 kilovolt-amperes) -- exclude wide-spread application.

Even so, there is every reason to expect that, with the development of improved scanning techniques and the development of new types of display elements, the problem of the television display panel will be solved. The discovery of the dynamic scattering mode in liquid crystals,[104] which makes it possible to convert a thin liquid layer from a transparent medium into a diffusely reflecting one by a very small expenditure of electric power, represents one promising avenue, permitting the use of ambient light for viewing the electrically generated picture.

CHAPTER V

Television Broadcasting

A. _Evolution_. Television broadcasting has, in
general, followed the pattern observed in radio broadcasting. In some countries -- in particular, in the
United States -- it has been entirely in the hands of
private enterprises. In others, such as Great Britain
and most of the countries of Western Europe, it is
controlled by government-chartered corporations with
well-defined sources of income and a large measure of
autonomy. Finally, in Russia and the countries of
Eastern Europe, television broadcasting is a state
function and supported directly by taxation like other
state functions. In several countries the pattern is
mixed -- thus, in Japan, Canada, and Brazil government
and privately operated stations exist side-by-side.

In the United States the costs of television broadcasting are met, in the main, by advertising revenue,
without direct expense to the individual set owner. A
group of stations forming the National Education Television Network (NET) forms an exception to this, being
financed by foundations and voluntary private contributions, matched, within limits, by federal funds. Many
of these stations also provide instructional television
services to schools and receive support for these from
school districts. Finally there are, in many areas,
cable television systems supported on a subscription
basis; in large part, these exist primarily to improve
the local quality of reception and do not originate
their own programs.

In countries in which government-chartered corporations are charged with television broadcasting, programming costs are usually met by license-fees from set
owners and/or taxes on the sale of sets. In some areas,
advertising provides a subsidiary source of revenue,
whereas in others it is specifically excluded from the
charter of the broadcasting corporation. In Great
Britain, two public corporations are charged with television broadcasting: The British Broadcasting Corporation (BBC), which has operated television stations
since the inception of public television broadcasting
in Great Britain in 1936, and the Independent Television
Authority (ITA), which was chartered in 1954. Whereas
BBC's programs are supported by license fees, ITA's are

sponsored and paid for by advertisers. The two systems
are competing aggressively for the television audience,
just as the major networks in the United States. In
1966, the ratio of the viewing time of ITA and BBC programs was 56:44.[105]

Some measure of public control of television broadcasting exists in all countries. This is a simple
consequence of the fact that some authority must assign
broadcasting channels to prevent interference which would
ruin reception for set-owners generally and limit other
public services. In the United States this authority
is exercised by the Federal Communications Commission.
In issuing licenses, the Commission is guided by the
principle that broadcasting stations must operate "in
the public interest, convenience, and necessity."[106]
Court decisions have upheld the right of the Commission
to consider, in granting and renewing licenses, the past
practices of the petitioner with respect to the public
interest as well as his technical competence to make
proper use of the channel assignment. While the position
has been taken that the Federal Communications Commission
should demand a reasonable distribution of broadcast
time over different areas of public interest, the
licensing power has generally been used only to prevent
the establishment of monopolies in news dissemination;
to deter gross breaches of good taste; to assure fairness in giving voice to opposing sides; and to avoid
gross violation of public interest as by medical mal-
practice over the air. The broadcasting industry through
the National Association of Broadcasters (NAB), has
attempted to police itself in matters such as the fraction of broadcast time which may be devoted to commercials although the effectiveness of this approach is
limited by the fact that many of the smaller stations
do not subscribe to the NAB code. Primarily, the
question of program content is left to the broadcast
corporation and the advertising sponsor.

The disinclination to have program content dictated
by the government, which characterizes the American
broadcasting system, is equally evident where public
corporations are responsible for broadcasting. Here,
too, we find insistence on fairness in the presentation
of opposing views and on full and accurate reporting --
a practice which is, without doubt, responsible for the
high international repute of e.g. the British Broadcasting Corporation. Even where broadcasting is entirely
a government function the power of the switch -- the
tendency of the viewer to turn off dull fare -- exercises

pressures for varied and interesting programs. This is
especially true in overlapping reception areas of
different television systems.

Irrespective of the nature of broadcast sponsorship,
television, throughout the world, has been enormously
successful in building up an audience. In the decade
1960-69 world receiver ownership has grown from 94
million to 225 million sets (Fig. 29) and the number
of television stations has grown from 2000 to 9000
(Fig. 30). Public opinion polls conducted in the United
States indicate, furthermore, that there has been a
steady growth in the acceptance of, and reliance on,
television as a medium of information:[107] By 1967,
over twice as many respondents in the general population
were inclined to keep television as the next most popular
medium (newspapers), if they were obliged to choose a
single one, being presented with television, newspapers,
radio and magazines as alternatives. A comparable
plurality regarded television as the most reliable source
of information. Furthermore, although the preference
for television was less pronounced among the upper
economic brackets and, particularly, among the college-
educated, the continuous growth in its acceptance cut
across the entire spectrum of the population. A further
measure of the impact of television in the United States
is given by the estimated average daily viewing time
(in 1967) of 2 hours and 41 minutes.

In summary, within the brief span of a quarter
century since television emerged from the laboratory,
it has become the most potent single cultural and
educational force which the world has ever known.

B. _Programming_. The character of television
programming depends in large measure on its financial
support. Advertising-supported television tends to
stress entertainment, whereas tax-supported television
places more emphasis on public education. Under author-
itarian rule, this is likely to have a substantial
admixture of political indoctrination.

In the United States the selection and continuance
of programs by commercial stations is based primarily
on audience response, for which sophisticated methods
of measurement and rating have been established; these
ratings measure in effect the exposure which the advertiser achieves for his message in sponsoring the
program. At first sight this might also seem the ideal
way of determining program content: through its

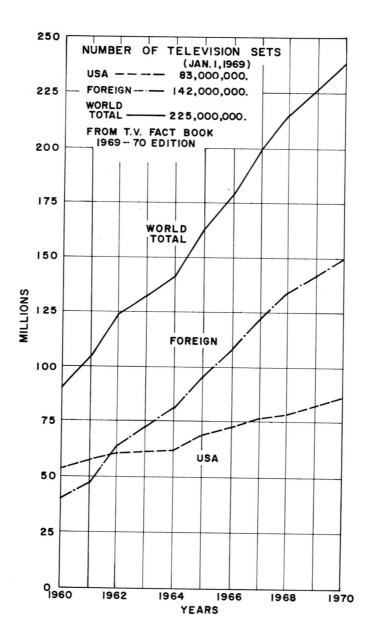

Fig. 29

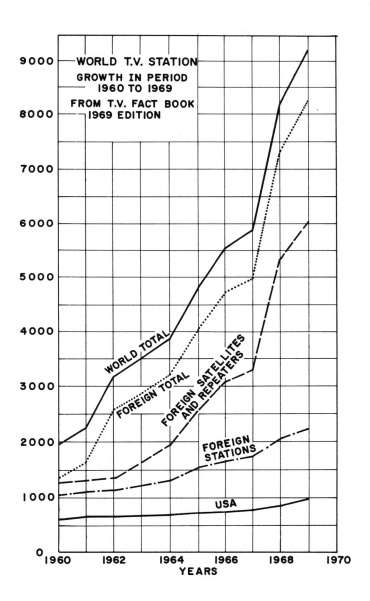

Fig. 30

manipulation of the station selector and the switch the
public dictates what programs it wishes to see and hear.
On the other hand, people may well watch low-calibre
programs even knowing that they derive no benefit from
them.   Parents, in particular, even though they may
defend their rights as adults to make such choices, will
rarely advocate an equal freedom for their children.
Pressure has thus been brought, in particular, to limit
the extent of violence portrayed in television programs
directed toward children and the major networks have
accepted responsibility for policing their presentations
in this respect.

A second difficulty with program selection based
on audience rating is the tendency of the major stations
in a given area to present programs of similar character
at the same time, each competing for a maximum slice
of the audience by the means which the rating system
has indicated to be most effective.   This greatly reduces the choice in program material available to the
television viewer.   In this respect, license-fee
supported television in the hands of government-chartered
corporations affords a better opportunity for planning
well-balanced broadcast schedules to meet public needs.

The shortcomings of advertising-supported television in the United States have been reduced by a
number of factors.   One of these is the presentation of
many excellent public-service programs by the networks.
Another is the support of programs of high public worth
irrespective of the rating by individual sponsors.
Finally the gradual development of National Educational
Television, free of advertising, has introduced a new
form of competition which will inevitably have a salutary
effect on program content.   Cable Television, which, in
1969, served 3,600,000 viewers in the United States,[108]
can also be expected to broaden the program choices for
those who can afford the subscription price.   Similar
cable television services are anticipated in Great
Britain.

Programs of broad public interest have been distributed, almost since the beginning of television
service, over the North American continent by cable
and microwave networks, subsequently, through the
Eurovision network, over much of the European continent,
and now, since 1965, by way of "Early Bird" satellites,
on an intercontinental basis.   The first manned landing
on the moon was viewed in all parts of the world as it
was occurring, prompting a remark by the President of

the United States to the effect that "for a time, mankind was one."

Apart from this simultaneous viewing, throughout the
world, of momentous events, the great appetite for program material of the many television stations scattered
over the Earth's surface is satisfied by the import of
tapes and films in vast quantities from the principal
originating centers in the United States, Great Britain,
and France. If this is used for distributing the best
that our cultures have to offer and not as a source of
revenue from second-rate material, it presents a great
opportunity in building a valid common heritage, quite
compatible with appreciation of indigenous culture and
achievement.

Public opinion polls suggest general satisfaction
with the present system of free, advertising-supported
television in the United States, and no desire for increased government regulation. At the same time, the
most broadly endorsed statement in the same polls[107]
(by 80% of the respondents), was to the effect that there
were far too many commercials on television. The persons
polled were also asked which of the following two statements most closely represented their own point of views:

"(Broadcast Television) serves the public interest..
a balanced blend of light and high-brow entertainment,
public affairs, and news...offers something for every
one" (Robert Sarnoff, 1959), and

"(Broadcast Television is) a vast wasteland...a
procession of game shows, violence, audience participation shows, formula comedies...private eyes, gangsters,
more violence, and cartoons..." (Newton Minow, 1961).

The respondents consistently showed a preference for
the first statement by a ratio which rose from 52:23 in
1961 to 62:20 in 1964 and then dropped again to 53:29
in 1967. The later decline suggests an increasingly
critical attitude by the public.

In democratically governed countries with fee- or
tax-supported television we can infer a comparable
support of their system from the fact that there has
been no great rush to shift the financial base of
television support to advertising.

The existence of general approval of the prevailing
system is not inconsistent with vehement criticism of

individual features of the system. Such criticism comes
from both the left and the right. In the United States
we find the Federal Communications Commission under
pressure, on the one hand, to insist on greater concern
with the public interest -- eventually at the cost of
advertising profits -- as a price for the issuance of a
broadcasting license. On the other, we find conservative
elements clamoring for the revocation of licenses from
stations which, in the view of the petitioners, have
gone too far in challenging traditional values. To a
considerable extent, the criticism of the left is
directed against the practices of commercial stations,
that of the right, against those of stations maintained
by voluntary subscription and foundation support. We
find a somewhat similar pattern in Great Britain,
although here the advertiser, as sponsor of the television programs of the ITA, is under much stricter con-
straints than in the United States. Even so, advertising-supported programs tend to be more conservative
in content and find their main response in the con-
servatively-minded bulk of the population, whereas the
license-fee supported BBC programs have been more
exploratory and daring and have received a corresponding
measure of conservative criticism.[105]

C. The Present Role of Broadcast Television. At
this time we find in the United States, and in many
other parts of the world, a salutary competition between different modes of dispensing television entertainment and instruction, with a variety of sponsorship.
This situation is furthered, and in turn encourages,
an increasingly discriminating attitude of the public
with respect to program material. The ever-increasing
facilities for program exchanges and the familiarity
of greater numbers of individuals with different
systems can be expected to stimulate similar critical
attitudes with respect to program selection where
broadcasting is under direct government control.

In such a competitive atmosphere a reasonable
limitation of commercial air time and imaginative
presentation of commercials, along with a consideration
of public interest in the selection of program material,
becomes essential for the survival of advertising-
supported television; similarly, the fee- or tax-supported broadcaster must reckon with the large public
demand for good entertainment, apart from instruction.
Fortunately, there is no basic conflict between the
claims of entertainment and instruction. Education,
at its best, holds its audience as effectively as the

as the best of entertainment and the latter, in turn,
has never been devoid of educational content.

Already, broadcast television has been used effectively by imaginative teachers to supplement the teaching
process.[109] The group witnessing of historic events on
the television screen can stimulate student discussion
and involvement in the problems of our day more effectively than any technique that has been available to us
in the past. Furthermore, just as broadcast television
can greatly help the teacher in his educational task,
the teacher can help the student in deriving the maximum
benefit from television by developing critical attitudes
in program selection.

There is thus every reason for optimism with respect
to the maturing of broadcast television as a major
factor in enriching life and broadening understanding
throughout the world. As with other major endeavors,
the time scale of the process of reaching this goal
will depend largely on the extent of citizen involvement.

CHAPTER VI

Other Uses of Television

      Although developed with entertainment broadcasting
as its primary function, television has many applications
in many other fields, some of which are discussed below.

      A. Industry. To meet the unprecedented demand
of the general public for TV receiving sets, the electronics industry began building specialized factories,
distribution facilities, and research laboratories.
This development, in turn, attracted engineers and
scientists from other branches of industry, and universities began shifting their enrollments to accomodate
students interested in this new engineering specialization.

      B. Education. The use of TV for educational
purposes is obvious. Courses, hitherto available only
at scattered locations, can now be given anywhere and
anytime - indeed, the early morning schedules of many
commercial television stations are replete with all
kinds of college courses, many often given for credit.
And of course, the primary aim of educational television
is inherent in its very name. Within universities and
primary schools, television is already a necessary
adjunct to teaching, and the use of taped lectures is
creating a permanent library of all kinds of knowledge.
Other institutions are also profiting by the use of TV
for training purposes. Unquestionably, these trends
will continue as the huge body of knowledge accumulated
by these methods is made available to more of the
worlds' populations.110*

      C. Politics. Television has had a profound effect
on political processes world-wide. In those countries
with representative forms of government, it provides
politicians easy access to populations to present their

_____

*As this article was being written, representatives of
the Ivory Coast, France, and three United Nations agencies formally agreed to collaborate on a 500-million
dollar project that would equip primary schools for a
million Ivorian children with television sets. Similar
instructional television systems were introduced in the
mid-1960's in American Samoa and, on an experimental
basis, in the West African republic of Niger.

often opposing points of view, leaving it up to the
viewer to make his own decision. The proximity of the
living human being offering his opinions has radically
altered the language of politics. In the United States,
the 1960 television debates of Kennedy and Nixon were
believed to be the decisive influencing factor in the
outcome of the election. In totalitarian countries,
television is an important adjunct to the political control of the population's thinking. As yet undetermined
in the socio-political realm are the effects of the
daily television viewing of specific events, such as
riots and wars, and the offered possibility that television could indeed be a causative factor.

D. <u>Military</u>. As stated before, the industrial
realization of TV's commercial potential could be traced
to the decade just before the second World War. Although
the war practically stopped this activity, the existing
body of television research was immediately utilized by
the military demand for electronics, as, for example,
in radar where the technology developed for production
of kinescopes was utilized for radar tubes. The expanded interest in short waves necessary for TV broadcasting was reflected in new tubes and specialized
circuits. The military also first utilized many special
electronic devices such as electron multipliers. In
many cases, however, paradoxically, these devices were
utilized for different purposes than what was originally
intended. The electron multiplier was invented as a
noiseless electron amplifier, but its first military
use was as a noise generator for radar countermeasures.
The sensitivity to invisible radiation, such as infrared,
of devices developed in television was used to facilitate
night operations. When the War was over, industries in
many countries were fully prepared to immediately begin
large-scale manufacturing of electronic receivers and
associated equipment, and thus it became possible to
distribute them for relatively modest prices due to the
mass production.

E. <u>Medicine</u>. The growth of the electronics industry and its subsequent attraction and creation of
new specialists and facilities could not fail to spread
its influence into other fields, such as medicine.
Miniaturization, which became very important with the
increased complexity of electronic circuits and the use
of higher frequencies for communication, was responsible
for the creation of new amplifying tubes and other
elements of the circuits. It revolutionized the field
by introducing transistors, which in turn converted the

whole electronics into a solid-state field.

Many new sensing devices in medicine became practical with miniaturized elements, i.e. heart pacers, radio
pills, and other nondestructive medical sensors became
standard items in medicinal care.

F. Computers. The development of electronic computers also profited from this newly developed electronic
industry. It is sufficient to compare the first operating electronic computer "Eniac" (which used twenty-
thousand radio tubes at hundreds of kilowatts of air-
cooling power) with the present desk computer of the
size of a typewriter, which in addition gives better
performance. The importance of this trend toward miniaturization becomes obvious.

CHAPTER VII

Indirect Contributions of Television

Industry and Research

Television has not only profoundly altered our
cultural environment but it has also added to our competence in dealing with countless problems by, in
effect, extending our senses. The process of bringing
it into being, both as a technical development and as
a world-wide business enterprise, has itself had important by-products, which we shall consider briefly.

A. Advancement of Solid-State Science and
Technology. The simple fact that television created an
enormous demand for devices the production of which required a high level of technological sophistication has
itself had a powerful effect on the advancement of
technology. The crucial event in the development of
solid state technology, the discovery of the transistor,[111] coincided with the development of a mass market
for television receivers. The awareness of potential
utilization for this mass market provided a powerful
incentive for broadening the understanding of solid-state
phenomena and the search for and perfection of new
solid-state devices. Industry was quite ready to provide funds for research and development which could
lead to economies or improvements in performance of
television equipment and thus provide an advantage in a
huge and highly competitive market. With such support
the lead-time between the conception of an idea or the
discovery of a new phenomenon and its use in production
or translation into a marketable device became short.
We have witnessed this course of events in the development of the transistor itself in its many variations;
in the evolution of integrated-circuit technology; in
the creation of new types of phosphors and electroluminescent materials; in many phases of circuit design,
which have left their impact on the electronic computer
field; and even in the advancement of lasers and holography. The existence of the television industry has
been, and continues to be, a powerful stimulant to the
advancement of the entire field of electronics and
optics.

B. Development of New Tools for Research. Long
before a television industry existed, studies undertaken
to create a technical basis for modern television led to

new devices which have become research tools of major
importance in many lines of investigation.

     C. The Multiplier Phototube. In our discussion of
the development of television camera tubes we noted that
one performance-limiting factor was the noise, or the
random variation in background level, injected by the
thermionic amplifier for the video signal derived from
the camera tube. In the image orthicon and other
high-sensitivity tubes, this noise source was effectively
eliminated by the insertion of a secondary-emission
multiplier for the return beam from the target, providing an essentially noise-free first stage of amplification.

     The first high-gain electron multiplier operated
with static fields employed a transverse magnetic field
to guide the electrons from one stage of secondary-
emission multiplication to the next.[112] This was soon
followed by a purely electrostatic type, in which the
successive target electrodes were shaped to guide the
secondary electrons from one target to that portion
of the next which was most favorable for further multiplication.[113] Both had their origin in the effort to
minimize the noise level in the transmitted television
picture.

     The electrostatic multiplier phototube, in par-
ticular, soon received recognition as the most sensitive
light detector in existence. With high sensitivity it
combined high speed of reaction, eventually being found
capable of resolving events a billionth of a second
apart. The scintillation counter,[114] combining a suitable phosphor with a multiplier phototube, became thus
an essential tool of nuclear physics, being capable of
detecting individual high-energy particles incident on
the scintillation phosphor by the light released within
the latter. The high time-resolution of the multiplier
phototube permitted, furthermore, the construction of
coincidence counters which could detect and count pairs
(or groups) or particles originating in the same nuclear
reaction.

     We have characterized the amplification by the
secondary emission multiplier as essentially noisefree.
Actually, our early studies of the electron multiplier
showed that the small amount of noise injected by the
amplification should be approximately inversely proportional to the gain of the first stage of multiplication. By using a material with a very high secondary-

emission gain for the first target electrode of a
scintillation-counter electron multiplier; it became
possible to discriminate small differences in the light
excitation in the phosphor, and hence in the energy of
the incident particle. Such very-high-gain materials
became available to us, as a result of fuller understanding of solid-state physics, in the gallium-phosphide
dynode with an adsorbed cesium layer.[42] With them, the
scintillation counter became capable of not only the
detection, but also the accurate measurement of the
energy of the incident particle and the statistics of
photoemission in response to its incidence.[115]

The multiplier phototube, in view of its great
sensitivity, has found many other important uses in
science and technology, such as light sensor in spectroscopy, as a detector of laser signals, in the study of
quantum statistics, etc. It is one of the truly in-
dispensable tools of modern science.

D. The Image Tube. The image tube was similarly
developed for increasing the sensitivity of television
camera tubes. In the image iconoscope and the image
orthicon it constituted the image section, which converted the light image formed at a photocathode into a
photoelectron image at the storage target. By accelerating the photo-electrons to a high potential, the energy
of the photoelectron image could be made a thousand
times that of the incident light image. If the photoelectron image was formed on a phosphor screen closely
coupled optically with the photocathode of a second
imaging stage, a substantial gain could be realized in
the photoelectron output (and hence in the brightness
of the final image) in spite of unavoidable energy conversion losses. Thus the material gain in light output
realized with a single-stage image tube could be increased materially in multi-stage image tubes.[116]
Alternatively, high gains were realized in image multipliers, in which secondary emission amplification gain
take place at successive transmission dynodes.[117]

The image tube thus had the capability of converting a light image into one of greater brightness.
Since the photocathode could be made to be sensitive
in the near infra-red and ultraviolet, beyond the
sensitivity range of the human eye, it had the further
capability of making infrared and ultraviolet images
visible.

Our first non-television application of the image
tube was in an infrared microscope, facilitating the
study of e.g. insect structures hidden by the chitinous
material opaque in the visible, but transparent to
infra-red radiation.[118] During the war years infrared-
sensitive image tubes incorporated in "snooperscopes"
and "sniperscopes" served the quite different purpose of
nocturnal warfare.[119] Image tubes became once more important to microscopy as converters of ultraviolet
radiation in the "Ultrascope" making the examination of
biological material in the ultraviolet almost as simple
as with visible radiation.[120] One of the most extensive
uses of image tubes has been as "image intensifiers"
for x-rays, reducing the x-ray exposure of the patient
in fluoroscopy and avoiding the necessity of dark
adaptation on the part of the surgeon or radiologist.[121]
The use for this purpose of the image intensifier
optically coupled to a television system has already
been noted. Image tubes have also had important applications in increasing the sensitivity of astronomical
cameras[122] and in electronically shuttered ultra-high-
speed photography.[123]

    E. The Electron Microscope. The first practical
application of the concept of an electron lens, in the
form of an axially symmetric electrostatic or magnetic
field, was to the convergence of an electron beam in a
high-vacuum cathode-ray tube into a small, sharply
defined, spot on the screen. This use was realized in
the earliest versions of the kinescope. In the same
tube the beam was deflected across the screen by a
transverse magnetic field, the functional analog of a
prism.

    The very term "electron lens" suggested the con-
struction of other electron analogs of optical instruments, such as an electron microscope. It is thus not
surprising that we find Knoll and Ruska,[124] at the
Technical University in Berlin, in 1931, engaged in the
construction of an electron microscope as a direct out-
growth of their work on high-voltage oscillographs. As
the main pressure in the development of television
shifted from the laboratory to the factory, we too
started work on an electron microscope, although there
was no known market for such a device at that time. Our
early work was largely directed toward the development
of a scanning electron microscope, which can indeed be
regarded as a specialized closed-circuit television
system.[125] Here the object is scanned by a very small

electron spot in a scanning pattern of greatly reduced
dimensions and the image is reproduced on the screen of
a cathode ray tube or by a facsimile printer. The mag-
nification is simply the ratio of the linear dimensions
of the image scanning pattern and the object scanning
pattern and the resolution is limited by the sharpness
of the scanning spot on the object.

    In more recent years the scanning electron micro-
scope has been developed into an instrument of great
power and versatility, equally valuable in the biological
sciences and the solid-state field.[126] However, our
early work did not achieve results exceeding those
realized by conventional optical methods by a sufficient
margin to justify the greater instrumental complexity.
Hence we concentrated our attention on a transmission
electron microscope, which soon demonstrated a resolution
capability far in excess of the conventional light
microscope.[127] In fact, the vast range of detail revealed by the new instrument, far beyond that known to micro-
scopists at the time, made interpretation difficult and
proved a hindrance to its acceptance -- along with the
need for developing a whole new technique of preparing
and handling specimens. However, while much inventive-
ness and diligent effort were required to overcome these
barriers, it was only a matter of time until investigations probing into the deepest secrets of nature,
down to molecular dimensions, availed themselves quite
generally of the electron microscope. Few of the users
of this powerful tool of science are likely to be aware
of the intimate connections of its origins with the
development of television.

CHAPTER VIII

Television: Its Role in the World - Today and Tomorrow

Predicting where new developments will lead is
always difficult and uncertain; what might seen to be
an obvious and close development could take decades and,
conversely, what might seem to be a far-off realization
could be announced in tomorrow's newspaper. However,
possibly looking at what already has been achieved and
what is yet dreamed about might give some clues.
Obviously, the enormous vitality and progress that has
already been achieved will continue, creating better
and more sophisticated devices and facilities than
are already in evidence.

The past history of television development indicates
that its phenomenal growth was made possible by its
acceptance for home entertainment. It is possible,
therefore, to visualize that the next mass appeal of
science to the multitude will come when some new development forcibly attracts their attention.

The fact that present integrated-circuit technology
can transform the "rat's nest" of a complex electronic
circuit into an orderly pattern resembling a color print
indicates the changes in television receiver design
which we can anticipate. In many respects, our present
situation resembles the transition period between the
mechanical and the electronic period of television which
was described in the first chapter. Now, as then, we
have at our disposal all the elements necessary to effect
the change. Technically, a printed television set,
without even the familiar cathode-ray picture tube or
kinescope, is a possibility, as has been pointed out in
Chapter IV. The only question is how soon, if ever,
such a "printed" set will be able to compete, in
quality and price, with conventional sets manufactured
by highly developed production techniques.

With an approach to a completely printed television
set, we can imagine a great simplification in the
servicing of television receivers. Thus, with the continuous broadening of the uses of the telephone and
with the receiver circuits in the form of standardized
circuit panels, it should become possible for local
service centers to perform routine computer-operated
tests of the set circuits over the telephone line,

sending out replacement panels for installation by the
set owner, with substantial savings in time and expense.

In the case of TV broadcasting, which is essentially
a means to contact a multitude instantly from a center,
it is possible to visualize a reverse system whereby
the multitude can communicate quickly with the center,
either in response to a proposal or question from the
center or to communicate their opinions on certain
important matters. If this system could be inexpensive,
would not demand much time or effort, and not overload
the existing communication network, it could be
accepted as universal for many other purposes that are
already in great demand now, i.e. consumer credit cards,
bank operation without cash transactions, etc. Without
discussing the technical details, it is possible to state
that the present electronics industry and the present
status of the communication art are perfectly adequate
to produce such a system (Telecard).*

Another example is the role of electronics in home
management. This is, of course, an extension of the
present, fast-growing use of electronic gadgets, but
the future will find much broader uses than those which
are current. For instance, TV can be used for the
observation by the housewife of small children,[†] home
appliances,[†] kitchen process,[†] visitors at the door,[†]
etc. Electronics can also provide a mechanical memory
for routine chores, send letters without writing or
telephoning, or order purchases and credit them to the
customer's account.

The extension of electronics for do-it-yourself
medical reporting may also be visualized. At present,
many individuals see doctors only when they feel it is
necessary, and sometimes without sufficient reasons,
thereby overloading the medical facilities.

It is possible to imagine a set of questions and
tests with simplified medical sensors recorded by

---

*A system first proposed by the senior author in
September 1969.

[†]First proposed by the senior author in 1954 (reference
10). As this paper was being written (April 1970) the
New York Times began carrying an advertisement from a
large New York Department store offering such a system
for sale to the public.

existing methods and transmitted by time sharing over
the telephone or picturephone to a medical center in
the individual's community. This information will be
analyzed by computer, and compared to the individual's
existing records for the most dangerous and obvious
symptoms. A clean bill of health may result, with the
alternative of a telephone appointment with a medical
staff member to inquire into additional symptoms or the
absence of specific symptoms. This, in turn, may lead
to a confirmation of the good health of the individual,
an appointment to see a proper specialist, or the
sending of an ambulance to his home.

Still another and fast growing necessity in our
present everyday life is improving the safety of our
roads. It is obvious that our life and safety depend
greatly on our transportation facilities, i.e., on roads
and vehicles. Even if we forget for a moment the air
pollution from automobiles which, today, is at the
center of public attention, the growing traffic jams
and mortality from automobile accidents require some
remedy. Obviously, traffic accidents will continue and
grow if driving is left entirely to individuals. Fa-
tigue, temporary loss of attention, or misleading road
signs all can contribute to accidents.

Electronics could be of great help here. In the
past few years, several proposals have been offered to
remedy these hazards. The feasibility of driving the
car automatically, i.e., keeping the car on the center
of the road, controlling the permissible range of speed
in a given sector, and measuring and controlling the
minimum distance from the car ahead, have all been
demonstrated over 10 years ago.[128]

One proposal visualized a gradual transition from
present driving practices. As a first step, an
electrical cable buried in the center of roads would
supply necessary visual or acoustic information to the
driver, without modifying present cars. As a next
step, cars would be equipped with automatic guidance
devices, which could be inactivated by the driver to
provide manual control. The introduction of new and
completely automatic cars would be the final step.
This system would permit the composing and recording
on tape of the planned trip, marking the route numbers,
mileages, and the turns that have to be made. Manual
driving would be necessary only at the beginning and
the end of the trip. It has even been suggested that a
way to avoid unnecessary trips would be to attend

parties by picturephone and to plan dinners by ordering
food to be delivered to guests by telephone from local
caterers.

We know from personal experience how we are con-
stantly being deluged by all kinds of written materials,
having neither adequate time for reading or adequate
space for storage, if they are read. For such a situation, electronics could be of enormous help. Instead
of written copies of books, journals, or even newspapers, we could receive by cable TV, on demand, titles
and abstracts. We could then select those we want and
order them delivered by TV or recorded in our home by
facsimile at the most convenient time. Just imagine
how much space could be saved and what enormous amount
of knowledge could be available to us without ever
leaving our homes!

We must admit that most of what we have discussed
in this last chapter is still only a dream. However,
these dreams are based on recent accomplishments, and
therefore, the fulfillment of such dreams requires
only the desire that they materialize. Thus, we have
tried to show that what we have now is the fruit of the
dreams of long ago. When we consider the rate at which
the time scale of development has been speeded up in
this century the realization of dreams of today may not
be too distant.

CHAPTER IX

References

1. N. W. Thomas, Crystal Gazing, Dodge, New York, 1905.

2. J. V. L. Hogan, "The Early Days of Television," J. Soc. Motion Picture Television Engrs., Vol. 63, pp. 169-173, 1954.

3. A. Korn and B. Glatzel, Handbuch der Photoele- graphie und Telautographie, Nemnich, Leipzig, 1911.

4. W. E. Sawyer, "Seeing by Electricity," Sci. American, Vol. 42, p. 373, 1880.

5. D. Mihaly, Das Elektrische Fernsehen und das Tele- hor, Krayn, Berlin, 1926.

6. A. A. Campbell-Swinton, "The Possibilities of Tele- vision," Wireless World and Radio Review, pp. 51-56, April 9, 1924.

7. G. Goebel, "Das Fernsehen in Deutschland bis zum Jahre 1945," Archiv für das Post und Fernmeldewesen Vol. 5, pp. 259-393, 1953.

8. E. W. Engstrom, "A Study of Television Image Char- acteristics," Sect. I, Proc. I.R.E., Vol. 21, pp. 1631-1651, 1933, and Sect. II, Proc. I.R.E., Vol. 23, pp. 295-310, 1935.

9. P. Mertz and F. Gray, "Theory of Scanning," Bell System Tech. J., Vol. 13, pp. 464-515, 1934.

10. V. K. Zworykin and G. A. Morton, Television, 2nd Edition, Wiley, New York, 1954.

11. V. K. Zworykin, "Television with Cathode-Ray Tube for Receiver," Radio Eng., Vol. 9, pp. 38-41, December 1929.

12. V. K. Zworykin, "Description of an Experimental Television System and the Kinescope," Proc. I.R.E., Vol. 21, pp. 1655-1673, 1933.

13. H. Busch, "On the Operation of the Concentrating Coil in the Braun Tube," Arch. Elektrotech.,

Vol. 18, pp. 583-594, 1927.

14. A. Dinsdale, "Television by Cathode Ray. The New Farnsworth System," Wireless World, Vol. 28, pp. 286-288, 1931.

15. M. von Ardenne, "The Braun Tube as Television Receiver," Fernsehen, Vol. 1, pp. 193-202, 1930; Vol..2, pp. 65-68 and 173-178, 1931.

16. V. K. Zworykin, "The Iconoscope -- a Modern Version of the Electric Eye," Proc. I.R.E., Vol. 22, pp. 16-22, 1934.

17. P. T. Farnsworth, "Television by Electron Image Scanning," J. Franklin Inst., Vol. 218, pp. 411-444, 1934.

18. A. G. Jensen, "The Evolution of Modern Television," J. Soc. Motion Picture Television Engrs., Vol. 63, pp. 174-188, 1954.

19. G. R. M. Garratt and A. H. Mumford, "The History of Television," Proc. Inst. Elec. Engrs., Vol. 99, Part III-A, pp. 25-42, 1952.

20. M. Dieckmann and R. Hell, "Lichtelektrische Bildzerlegerrohre für Fernseher," German Patent 450,187, issued April 5, 1925.

21. See Farnsworth and U. S. Patent 1,773,980 of Jan. 7, 1927, and 1,986,330 of April 17, 1928.

22. P. K. Weimer, S. V. Forgue and R. R. Goodrich, "The Vidicon Photoconductive Camera Tube," Electronics, Vol. 23, pp. 70-73, May 1950; RCA Rev., Vol. 12, pp. 306-313, 1951.

23. H. Hertz, "Ultraviolet Light and Electric Discharge," Ann. Phys., Vol. 31, pp. 983-1000, 1887.

24. J. Elster and H. Geitel, "The Use of Sodium Amalgam in Photoelectric Experiments," Ann. Phys., Vol. 41, pp. 161-165, 1890.

25. J. Elster and H. Geitel, "Proportionality of Light Intensity and Photocurrent in Alkali Metal Cells," Physik. Z., Vol. 14, pp. 741-752, 1913.

26.   P. Lenard, "Production of Cathode Rays by Ultra-
      violet Light," Ann. Phys., Vol. 2, pp. 359-375,
      1900.

27.   J. J. Thomson, "On the Masses of the Ions in
      Gases at Low Pressures," Phil. Mag., Vol. 48,
      pp. 547-567, 1899.

28.   A. Einstein, "A Heuristic Standpoint Concerning
      the Production and Transformation of Light,"
      Ann. Phys., Vol. 17, pp. 132-148, 1905.

29.   M. Planck, "On the Theory of the Law of Energy
      Distribution in the Normal Spectrum," Verhandl.
      deut. physik. Ges., Vol. 2, pp. 237-245, 1900.

30.   L. R. Koller, "Photoelectric Emission from Thin
      Films of Cesium," Phys. Rev., Vol. 36, pp. 1640-
      1647, 1930.

31.   P. Goerlich, "On Composite Transparent Photo-
      cathodes," Z. Physik, Vol. 101, pp. 335-342, 1936.

32.   A. H. Sommer, U.S. Patent 2,285,062; A. H. Sommer
      and W. E. Spicer, "Bismuth-Silver-Oxygen-Cesium
      Photocathode," J. Appl. Phys., Vol. 32, pp. 1036-
      1042, 1961.

33.   A. H. Sommer, "New Photoemissive Cathodes of High
      Sensitivity," Rev. Sci. Instr., Vol. 26, pp. 725-
      726, 1955.

34.   J. J. Scheer and J. von Laar, "GaAs-Cs: A New
      Type of Photoemitter," Solid State Commun., Vol. 3,
      pp. 189-193, 1965.

35.   T. E. Fischer, "Photoelectric Emission and Inter-
      band Transitions in GaP," Phys. Rev., Vol. 147,
      pp. 603-607, 1966.

36.   L. Austin and H. Starke, "On the Reflection of
      Cathode Rays and a New Phenomenon of Secondary
      Emission Related to It," Ann. Physik, Vol. 9,
      p. 271, 1902.

37.   H. Bruining, "Physics and Applications of Secondary
      Electron Emission," McGraw-Hill, New York, 1954.

38.   K. G. McKay, "Secondary Electron Emission," Ad-
      vances in Electronics, Vol. 1, pp. 65-130, 1948.

39. V. K. Zworykin, J. E. Ruedy, and E. W. Pike,
     "Silver-Magnesium Alloy as a Secondary-Emitting
     Material," J. Appl. Phys., Vol. 12, pp. 696-698,
     1941.

40. D. L. Emberson, A. Todhill and W. L. Wilcock,
     "Further Work on Image Intensification with
     Transmitted Secondary-Electron Multiplication,"
     Advances in Electronics and Electron Physics, Vol.
     16, pp. 127-139, 1962.

41. G. W. Goetze, "Transmission Secondary Emission for
     Low-Density Deposits of Insulators," Advances in
     Electronics and Electron Physics, Vol. 16, pp.
     125-154, 1962.

42. R. E. Simon and B. F. Williams, "Secondary-
     Electron Emission," IEEE Trans. Nucl. Sci., Vol.
     NS-15, No. 3, pp. 167-170, 1968.

43. V. K. Zworykin and G. A. Morton, "Television," 2nd
     Ed., Wiley, New York, 1954, Chapter 9.

44. V. K. Zworykin, G. A. Morton, and L. Malter, "The
     Secondary Emission Multiplier - A New Electronic
     Device," Proc. Inst. Radio Engrs., Vol. 24,
     pp. 351-375, 1936.

45. H. Busch, "On the Operation of the Concentrating
     Coil in the Braun Tube," Arch. Elektrotech., Vol.
     18, pp. 583-594, 1927.

46. H. W. Leverenz, "An Introduction to Luminescence
     of Solids," Wiley, New York, 1950.

47. W. Phillips and Z. J. Kiss, "Photo-Erasable Dark
     Trace Cathode-Ray Storage Tube," Proc. IEEE, Vol.
     56, p. 2072, 1968.

48. A. H. Rosenthal, "A System of Large-Screen Tele-
     vision Reception and on Certain Phenomena in
     Crystals," Proc. I.R.E., Vol. 28, pp. 202-212,
     1940.

49. F. Fischer and H. Thiemann, "Theoretical Consider-
     ations on a New Process for Large-Screen Television
     Projection," Schweizer Archiv f. angew. Wiss. u.
     Technik, Nrs. 1 and 2, pp. 1-23, 1941.

50.   D. H. Pritchard, "A Reflex Electro-Optic Light
      Valve Television Display," RCA Review, Vol. 30,
      pp. 567-592, 1969.

51.   J. S. Donal, Jr., and D. B. Langmuir, "A Type of
      Light Valve for Television Reproduction," Proc.
      I.R.E., Vol. 31, pp. 208-214, 1943.

52.   G. H. Heilmeier, L. A. Zanoni and L. A. Barton,
      "Dynamic Scattering: A New Electrooptic Effect in
      Certain Classes of Nematic Liquid Crystals," Proc.
      IEEE, Vol. 56, pp. 1162-1171, 1968.

53.   J. van Raalte, "Reflective Liquid Crystal Tele-
      vision Display," Proc. IEEE, Vol. 56, pp. 2146-
      2149, 1968.

54.   Thomas Young, "Lectures on Natural Philosophy and
      Mechanical Arts," London. 1807.

55.   H. von Helmholtz, Ann. Physik, Vol. 87, pp. 45- ,
      1852.

56.   J. C. Maxwell, Proc. Roy. Inst. of Great Britain,
      Vol. 6, p. 260, 1871.

57.   P. K. Brown and G. Wald, "Visual Pigment in Single
      Rods and Cones of the Human Retina," Science,
      Vol. 144, pp. 45-52, 1964.

58.   W. B. Marks, W. H. Dobelle and E. F. MacNichol,
      Jr., "Visual Pigments of Single Primate Cones,"
      Science, Vol. 143, pp. 1181-1182, 1964.

59.   P. C. Goldmark, J. W. Christensen, and J. J.
      Reeves, "Color Television -- U.S.A. Standard,"
      Proc. I.R.E., Vol. 39, pp. 1288-1313, 1951.

60.   A. V. Bedford, "Mixed Highs in Color Television,"
      Proc. I.R.E., Vol. 38, pp. 1003-1009, 1950.

61.   G. H. Brown and D. G. C. Luck, "Principles and
      Development of Color Television Systems," RCA
      Review, Vol. 14, pp. 144-204, 1953.

62.   "NTSC Signal Specification," Proc. I.R.E., Vol. 42,
      pp. 17-19, 1954.

63.   V. K. Zworykin, G. A. Morton, and L. E. Flory,
      "Theory and Performance of the Iconoscope,"

Proc. I.R.E., Vol. 25, pp. 1071-1092, 1937.

64. H. Iams, G. A. Morton, and V. K. Zworykin, "The Image Iconoscope," Proc. I.R.E., Vol. 27, pp. 541-547, 1937.

65. A. Rose and H. Iams, "The Orthicon, a Television Pickup Tube," RCA Rev., Vol. 186-199, 1939.

66. A. Rose, P. K. Weimer, and H. B. Law, "The Image Orthicon -- A Sensitive Television Pickup Tube," Proc. I.R.E., Vol. 34, pp. 424-432, 1946.

67. P. K. Weimer, "The Image Isocon -- An Experimental Television Pickup Tube Based on the Scattering of Low Velocity Electrons," RCA Rev., Vol. 10, pp. 366-386, 1949.

68. P. K. Weimer, S. V. Forgue, and R. R. Goodrich, "The Vidicon Photoconductive Camera Tube," Electronics, Vol. 23, pp. 70-73, May, 1950.

69. A. G. van Doorn, "The 'Plumbicon' Compared with Other Television Camera Tubes," Philips Tech. Rev., Vol. 27, pp. 1-14, 1966.

70. M. H. Crowell, T. M. Buck, G. F. Labuda, J. V. Dalton and E. J. Walsh, "A Camera Tube with A Silicon Diode Array Target," Bell Syst. Tech. J., Vol. 46, pp. 491-495, 1967.

71. M. H. Mesner and M. G. Staton, "Television in Space," Signal, Vol. 24, Nr. 5, pp. 9-13, 1970.

72. M. J. Cantella, "The Return-Beam Vidicon," Applied Optics, Vol. 9, 1970 (in print).

73. L. T. Sachtleben, D. J. Parker, G. L. Allee and E. Kornstein, "Image Orthicon Color Television Camera Optical System," RCA Rev., Vol. 13, pp. 27-33, 1952.

74. D. W. Epstein, "Colorimetric Analysis of RCA Color Television System," RCA Rev., Vol. 14, pp. 227-258, 1953.

75. F. G. Bingley, "Colorimetry in Color Television," Proc. I.R.E., Vol. 41, pp. 838-851, 1953 and Vol. 42, pp. 48-57, 1954.

76. H. de Lang and G. Bouwhuis, "Colour Separation in Colour-Television Cameras," _Philips Tech. Rev._, Vol. 24, pp. 263-271, 1963.

77. C. J. Hirsch, "Four Tube Separate Luminance Color Television Camera," _Broadcast News_, Vol. 133, pp. 30-47, April 1967.

78. H. Breimer, W. Holm, and S. L. Tan, "A Color Television Camera with 'Plumbicon' Camera Tubes," _Philips Tech. Rev._, Vol. 28, pp. 336-351, 1967.

79. "4 New Low-Cost Color Cameras Widen ETV and CATV Horizons," _Broadcast News_, Vol. 141, pp. 46-47, March 1969.

80. P. K. Weimer, W. S. Pike, G. Sadasiv, F. V. Shall-cross, and L. Meray-Horvath, "Multielement Self-Scanned Mosaic Sensors," _IEEE Spectrum_, Vol. 6, pp. 52-65, March 1969.

81. V. K. Zworykin and G. A. Morton, "_Television_," Wiley, New York, 1954; see Chapter 11, "The Kinescope," pp. 383-442.

82. "Direct-View Color Kinescope," _Proc. I.R.E._, Vol. 39, pp. 1177-1230, 1951.

83. H. B. Law, "A Three-Gun Shadow-Mask Color Kinescope," _Proc. I.R.E._, Vol. 39, pp. 1186-1194, 1951.

84. D. S. Bond, F. H. Nicoll and D. G. Moore, "Development and Operation of a Line-Screen Color Kinescope," _Proc. I.R.E._, Vol. 39, pp. 1218-1230, 1951.

85. G. V. Barnett, F. J. Bingley, S. L. Parsons, G. W. Pratt and M. Sadowsky, "A Beam-Indexing Color Picture Tube -- The Apple Tube," _Proc. I.R.E._ Vol. 44, pp. 1115-1120, 1956.

86. R. Dressler, "The PDF Chromatron - A Single or Multi-Gun Tri-Color Cathode Ray Tube," _Proc. I.R.E._, Vol. 41, pp. 851-858, 1953.

87. D. G. Fink, "Phosphor Strip Tricolor Tubes," _Electronics_, Vol. 24, pp. 89-91, December 1951.

88. A. C. Schroeder, U.S. Patent 2,446,791 issued August 10, 1948.

89. E. G. Ramberg, H. B. Law, H. S. Allwine, D. C. Darling, C. W. Henderson and H. Rosenthal, "Focusing Grill Color Kinescopes," IRE Convention Record, Vol. 4, part 3, pp. 128-134, 1956.

90. I. G. Maloff and W. A. Tolson, "A Resume of the Technical Aspects of RCA Theatre Television," RCA Review, Vol. 6, pp. 5-11, 1941.

91. S. L. Bendell and W. J. Neely, "Medium-Screen Color Television Projection," J. Soc. Motion Pict. and TV Engrs., Vol. 67, pp. 166-168, 1958.

92. E. Baumann, "The Fischer Large-Screen Projection System," J. Brit. I.R.E., Vol. 12, pp. 69-78, 1952.

93. William E. Good, "A New Approach to Color Television Display and Color Selection Using a Sealed Light Valve," Proc. Nat'l Electronics Conf., Vol. 24, pp. 771-774, 1968.

94. Y. Yamada, "Laser Color Television: The Largest Color Television Picture in the World," Laser u. angew. Strahlentechnik, Vol. 1, Nr. 3, p. 17, 1969.

95. G. Goebel, "Television in Germany up to 1945," Arch. Post u. Fernmeldewesen, Vol. 5, pp. 259-393, 1953 (see p. 362).

96. H. Henisch, "Electroluminescence," Pergamon, N.Y., 1962.

97. A. G. Fischer, "Injection Electroluminescence," Solid State Electronics, Vol. 2, pp. 232-246, 1961.

98. B. M. Arora, D. L. Bitzer, H. G. Slottow and R. H. Willson, "The Plasma Display Panel - A new Device for Information Display and Storage," 8th Nat'l Symp. on Information Display, 1967.

99. J. A. Rajchman, G. R. Briggs and A. W. Lo, "Transfluxor-Controlled Electroluminescent Display Panels," Proc. I.R.E., Vol. 46, pp. 1808-1824, 1958.

100. T. N. Chin, "A Magnetostrictive Device for Use in Scanning," Proc. IEEE, Vol. 54, pp. 1216-1217, 1966.

101.   S. Yando, "A Solid-State Display Device," Proc.
       I.R.E., Vol. 50, pp. 2445-2451, 1962.

102.   E. A. Sack, "The 'Elf' Screen," Research, Vol. 12,
       pp. 54-60, 1959.

103.   S. Shimada, "Setting the Stage for Flat-Screen
       TV," Electronics, Vol. 41, pp. 93-101, April 15,
       1968.

104.   G. H. Heilmeier, L. A. Zanoni and L. A. Barton,
       "Dynamic Scattering: A New Electrooptic Effect in
       Certain Classes of Nematic Liquid Crystals,"
       Proc. IEEE, Vol. 56, pp. 1162-1171, 1968.

105.   BBC Handbook 1966, British Broadcasting Corpora-
       tion, London.

106.   J. E. Coons, "Freedom and Responsibility in Broad-
       casting," Northwestern University Press, 1961.

107.   B. W. Roper, "Emerging Profiles of Television and
       Other Mass Media: Public Attitudes 1959-67,"
       Television Information Office, New York.

108.   Official Transcript, 18th Annual NCTA Convention,
       San Francisco, June 22-25, 1969, p. 665.

109.   Gloria Kirschner, "Start Where the Child Is,"
       Elementary English, November, 1969.

110.   The New York Times, April 12, 1970.

111.   J. Bardeen and W. Brattain, Phys. Rev., Vol. 48,
       pp. 230-231, 1948.

112.   V. K. Zworykin, G. A. Morton, and L. Malter, "The
       Secondary-Emission Multiplier - A New Electronic
       Device," Proc. I.R.E., Vol. 24, pp. 351-375, 1936.

113.   V. K. Zworykin and J. A. Rajchman, "The Electro-
       static Electron Multiplier," Proc. I.R.E., Vol. 27,
       pp. 558-566, 1939.

114.   G. A. Morton, "The Scintillation Counter," Ad-
       vances in Electronics, Vol. 4, pp. 69-107, 1952.

115.   G. A. Morton, H. M. Smith, and H. R. Krall,
       "Pulse Height Resolution of High-Gain First Dynode
       Photomultipliers," Appl. Phys. Lett., Vol. 13,

pp. 1356-1357, 1968.

116.   R. G. Stoudenheimer, "Image Intensifier Develop-
       ments in the RCA Electron Tube Division," Adv.
       in Electronics and Electron Phys., Vol. 12, pp.
       41-57, 1960.

117.   D. L. Emberson, A. Todkill, and W. L. Wilcock,
       "Further Work on Image Intensifiers with Trans-
       mitted Secondary Electrons," Adv. In Electronics
       and Electron Phys., Vol. 16, pp. 127-139, 1962.

118.   V. K. Zworykin and G. A. Morton, "Applied Electron
       Optics," J. Opt. Soc. Am., Vol. 26, pp. 181-189,
       1936.

119.   G. A. Morton and L. E. Flory, "Infra-Red Image
       Tube," Electronics, Vol. 19, pp. 112-114, Sept.
       1946.

120.   "Ultraviolet Image Converter Tube," Electronics,
       Vol. 32, p. 78, Feb. 20, 1959.

121.   M. C. Teves and T. Tol, "Electron Intensification
       of Fluoroscopic Image," Philips Tech. Rev., Vol.
       14, pp. 33-34, 1952/53.

122.   M. F. Walker, "Recent Progress in the Use of the
       Lallemand Electronic Camera in Astronomical
       Spectroscopy," Adv. in Electronics and Electron
       Phys., Vol. 22B, pp. 761-780, 1966.

123.   R. G. Stoudenheimer and J. C. Moor, "An Image Con-
       verter Tube for High-Speed Photographic Shutter
       Service," RCA Rev., Vol. 18, pp. 322-331, 1957.

124.   M. Knoll and E. Ruska, "Electron Microscope,"
       Z. Physik, Vol. 86, pp. 448-450, 1932.

125.   V. K. Zworykin, J. Hillier and R. L. Snyder, "A
       Scanning Electron Microscope," ASTM Bulletin,
       No. 117, pp. 15-23, 1942.

126.   C. W. Oatley and K. C. A. Smith, "The Scanning
       Electron Microscope and Its Field of Application,"
       Brit. J. Appl. Phys., Vol. 6, pp. 391-399, 1955.

127.   V. K. Zworykin, J. Hillier, and A. W. Vance, "An
       Electron Microscope for Practical Laboratory
       Service," Elec. Eng., Vol. 60, pp. 157-161, 1941.

128.   V. K. Zworykin and L. E. Flory, "Electronic Con-
       trol of Motor Vehicles on the Highway," reprinted
       from Highway Research Board Proceedings, Vol. 37,
       1958.

BIOGRAPHICAL NOTES

John Bardeen wrote most of his essay during a one month stay at the Center for Theoretical Studies in 1969. His work on superconductivity which is described in the article helped him to achieve in 1972 the unusual distinction of being the only scientist ever to be awarded the Nobel Prize in Physics for a second time. It was shared with his younger colleagues Leon Cooper and J. Robert Schrieffer for the theoretical work they had done during the 1950's. Professor Bardeen's first Nobel Prize was given in 1956 to him and his colleagues William Shockley and Walter Brattein for the invention of the transistor during the 1940's. He therefore is largely responsible for two of the most interesting discoveries in physics that have had profound impacts on the technology and through the latter on other sectors of society.

Professor Bardeen was born in Wisconsin in 1908. He took his Ph.D. at Princeton University in 1936 under Eugene P. Wigner and then spent periods at Harvard University, University of Minnesota, Naval Ordnance Laboratory and Bell Telephone Laboratories. Since 1951 he has been Professor of Electrical Engineering and Physics at the University of Illinois and since 1959 a member of the Center for Advanced Study of the University. He has received many honors for his work in theoretical solid state physics and is an active member of a number of scientific organizations. Through it all he has remained a serious and unassuming person who is dedicated to his researches in science.

* * *

Manfred A. Biondi has had an active and productive career in many aspects of atomic physics, plasma physics and solid state physics.

He was born in New Jersey in 1924, and received his B.S. and Ph.D. from Massachusetts Institute of Technology. He has been associated professionally with MIT and Westinghouse Research Laboratory. Since 1960 he has been Professor of Physics at the University of Pittsburgh. He was at the Center for Theoretical Studies in 1971 for the specific purpose of preparing his essay.

P.A.M. Dirac was born in 1902 in Bristol, England. He received his Ph.D. from Cambridge University in 1926 and stayed there as Lucasian Professor of Mathematics from 1932 until his retirement in 1968. In the latter year he came to the Center for Theoretical Studies for a three year stay, during which time he gave the lecture on which his essay is based.

Professor Dirac is one of the principal architects of the quantum theory, and made several unique contributions to science, including the reconciliation of relativity and quantum theory that led to the discovery of the well known relativistic wave equation which bears his name, the prediction of the existence of antimatter, the prediction of the spin of the electron, formulation of statistics of fields and particles, the predictions of gravitational waves and of magnetic monopoles, and recent work on cosmology.

He has been called by C.P. Snow the greatest Englishman since Isaac Newton, and few will question his designation as one of the great scientists who ever lived. His early work, besides bringing him the Nobel Prize in 1933, was instrumental in opening the paths that led to perhaps a dozen subsequent Nobel Prizes. It is quite possible that the ultimate impact of some of his discoveries has not yet been realized.

Professor Dirac is known as a man, not only in physics but in ordinary conversations, who does not make trivial remarks. There is a definite deep and well defined meaning in his every sentence. He is endowed with all the great virtues of a great man; has no enmities, no dislikes for any human being. He is a free man in the true sense of the word and also very courageous. We believe that his essay is an important document of the human intellect and spirit.

* * *

Willis E. Lamb, Jr. has been a frequent visitor at the Center for Theoretical Studies since its founding and has been a member of its Scientific Council since 1972. He is a very shy and modest person, with a dry but twinkling sense of humor.

Professor Lamb includes in his essay a brief description of how close he came to discovering the maser

during his researches on the fine structure of hydrogen.
He was awarded the Nobel Prize in 1955 for the last
named work. This research was based on an extension of
Dirac's theory of the hydrogen atom. Professor Lamb
has done important experimental and theoretical work in
a large number of fields, including nuclear and atomic
structure, cosmic rays, neutron physics, beta decay,
microwave spectroscopy and magnetrons. In recent years
he has been very active in the study of lasers and related questions.

Professor Lamb was born in Los Angeles in 1913. He
was educated at the University of California, receiving
his Ph.D. in 1938, and has served as a faculty member
at Columbia University, Stanford University and Oxford
University. Since 1962 he has been Professor of Physics
at Yale University.

* * *

Arthur L. Schawlow is, regrettably, the only con-
tributor to this volume who has not been in residence
at the Center for Theoretical Studies. His essay is
nevertheless a valuable personal view of the evolution
of nonlinear optics.

Since his essay contains much autobiographical
material, we will primarily add that he is one of the
principal figures in the development of the laser. He
has been Professor of Physics at Stanford University
since 1961, where he has continued the research in
quantum electronics and lasers and spectroscopy which
he began at Columbia University and continued at Bell
Telephone Laboratories. He has been honored for his
important discoveries with a number of important prizes.
Professor Schawlow was born in 1921 in New York and was
educated at the University of Toronto, receiving his
Ph.D. in 1949.

* * *

Edward Teller, who was born in Budapest in 1908,
is one of that remarkable group of Hungarian scientists
who in exile have made such a strong impression on
recent scientific history.

Much of Professor Teller's career has been

chronicled in the press and books. He has received
much recognition for his scientific work in addition to
his well-known development of the fusion bomb. (He has
remarked at the Center that he does not wish to be
known as the father of anything except of his children).
Professor Teller has been a Fellow of and member of the
Scientific Council of the Center for Theoretical Studies
since 1968. He has been in residence at the Center on
many occasions, contributing to its research program,
lecturing to all levels of audiences (including a set
of brilliant expository talks to gifted high school
students).

Professor Teller received his Ph.D. at the
University of Leipzig in Germany in 1930 and subsequently spent a number of years at Universities in Europe
and the United States before settling at the University
of California in 1953. Besides his thermonuclear research he has made important theoretical discoveries in
beta decay, molecular spectroscopy, and nuclear physics.

* * *

Vladimir K. Zworykin must be considered as one of
the most important figures of the electronics revolution
that has changed the face of our globe. He is the inventor of the iconoscope, kinescope, the photomultiplier,
image tube and electron microscope and many other
devices. His primacy in the invention of the modern
television system is generally acknowledged. (He has
been quoted as saying that the best part about television is the switch). He has been honored for his
important work by the principal scientific and engineering societies of many countries. He has been a
Fellow of the Center for Theoretical Studies since 1969,
spending about four months of each year in residence.
Dr. Zworykin is still actively engaged in a number of
important projects which cover the spectrum of human
activity, e.g. medical electronics, controlled thermonuclear power sources, mass transportation schemes, etc.

He was born in 1889 in Mourom, Russia and received
an Electrical Engineering degree from the Petrograd
Institute of Technology in 1912. His professor there
was one of the television pioneers, Rosing. After
serving as a Signal Corps Officer in the Russian Army,
he came to Westinghouse Corporation as a research engineer. He was brought by David Sarnoff to RCA in 1929,

when they began their long and fruitful association.
He was Director of the Electronic Research Laboratories
from 1929 to 1954 and has been an Honorary Vice President
of RCA since 1954. Dr. Zworykin is a man of great personal charm who is held in deep affection by many people
from all walks of life and nations. His former colleagues
and assistants head many of the major corporations in
this country.